Relevant Characteristics of Power Lines Passing Through Urban Areas

Relevant Characteristics of Power Lines Passing Through Urban Areas

Ljubivoje M. Popović
School of Electrical Engineering,
University of Belgrade, Belgrade, Serbia

ELSEVIER

ACADEMIC PRESS
An imprint of Elsevier

Academic Press is an imprint of Elsevier
125 London Wall, London EC2Y 5AS, United Kingdom
525 B Street, Suite 1650, San Diego, CA 92101, United States
50 Hampshire Street, 5th Floor, Cambridge, MA 02139, United States
The Boulevard, Langford Lane, Kidlington, Oxford OX5 1GB, United Kingdom

Notices
Knowledge and best practice in this field are constantly changing. As new research and experience broaden our understanding, changes in research methods, professional practices, or medical treatment may become necessary.

Practitioners and researchers must always rely on their own experience and knowledge in evaluating and using any information, methods, compounds, or experiments described herein. In using such information or methods they should be mindful of their own safety and the safety of others, including parties for whom they have a professional responsibility.

To the fullest extent of the law, neither the Publisher nor the authors, contributors, or editors, assume any liability for any injury and/or damage to persons or property as a matter of products liability, negligence or otherwise, or from any use or operation of any methods, products, instructions, or ideas contained in the material herein.

British Library Cataloguing-in-Publication Data
A catalogue record for this book is available from the British Library

Library of Congress Cataloging-in-Publication Data
A catalog record for this book is available from the Library of Congress

ISBN: 978-0-323-91136-8

Publisher: Joseph P. Hayton
Acquisitions Editor: Lisa Reading
Editorial Project Manager: Alice Grant
Production Project Manager: Nirmala Arumugam
Cover Designer: Mark Rogers

Typeset by MPS Limited, Chennai, India

In memory of my early deceased parents,
Matija and Milena

Contents

Preface

This book presents all stages of the process of developing the methodology that enables the solution of a problem long known in the professional and scientific public in the field of electric power distribution. It is about the electromagnetic influence of power lines on metal installations that surround these lines in urban conditions and circumstances, as well as the feedback effect of these installations on those lines. So, from the point of view of the theory of electric circuits, it can be said that it is a problem of inductively coupled lines with a common return path through the earth that was solved a long time ago. However, in urban areas, the problem arises because the mentioned installations are mostly placed underground and as such are inaccessible for collecting the data needed to calculate their own and mutual impedances as well as to measure the currents that occur in them. As a result, this problem for a long time was considered in professional and scientific circles as practically unsolvable.

Since along and around each electric power line, a fluctuating magnetic field is created spontaneously, it is not difficult to conclude that this problem appeared in practice simultaneously with the first applications of Tesla's alternative currents, that is, at the outset of the 20th century. Also, the problem in principle becomes solvable in the second half of the thirties of the last century, after the appearance of Carson's equations, that is, the analytical expressions that enable calculating the self- and mutual impedances of inductively coupled power lines and their conductors. However, since it is practically impossible to collect all data necessary for the calculation of these impedances, the problem of determining the electromagnetic interaction between power lines and different, known, and unknown metal installations typical of urban and suburban areas, remained unsolved until recently. Due to this the characteristics of power lines relevant for their planning, designing, and utilizing in urban and suburban areas have been determined long time with no taking into account the favorable inductive influence of the surrounding metal installations, that is, under the unrealistic assumption that these installations do not exist.

The here-presented methodology enables the compensation of the deficiency of all relevant but unknown data about the surrounding metal installations. This is achieved through the test measurements of their cumulative effect upon the value of the test currents obtained in the two of phase conductors by a simulated ground fault in the supplied substation and by using one of these conductors as one of the neutral conductors of the considered power line. The analytical part of the method is based on the introduction of a fictitious conductor which is equivalent, from the standpoint of the inductive effect for all known and unknown surrounding metal installations. The physical appearance of this imagined conductor is such that it represents a cylinder surrounding the conductors of the considered power line along their entire length. In the case of determination of inductive influence to one separately seen metal installation exposed only along one section to the considered power line, this fictitious conductor is also of cylindrical shape but surrounds only the considered metal installation along the section of exposure.

Also, the here-presented methodology can be applied to each, medium voltage, high voltage (HV), and extra-high voltage distribution line constructed in any urban or suburban environment. With small modifications the methodology can be used to determine the actual distribution of the ground-fault current, the actual inductive influence of power lines on any of the nearby metal installations, and the actual series impedances of power distribution lines. The results of the quantitative analysis show that the real picture concerning the magnitude of the grounding problem of HV substations located in urban areas is far more favorable than it was considered before. Also, the inductive influence of power lines upon surrounding metal installations, seen individually, is significantly smaller than it was possible to know before. Finally, as most important from the point of view of economical effects, it has been shown that the increase of transmission capacity of the existing power lines is considerable when we take into account the favorable inductive influence of the surrounding metal installations.

The book is assigned, first of all, for the professionals involved in the planning, design, and exploitation of lines and substations in power distribution networks, but, owing to the development of the original methodology with a completely new approach in solving the mentioned problems, it has a wider significance. For this reason, it can also be recommended to the academic people: professors, postgraduates, graduates, and students of electrical engineering.

Ljubivoje M. Popović

Introductory considerations

1

1.1 Problem description

Modern trends in the planning and development of electric-power systems are primarily orientated toward as much as possible better and more rational utilization of existing resources of these systems with the aim of achieving as much as more efficient power delivery. In the last more than two decades, these trends are known under the common name "smart grid." This concept imagined as perfect utilization of the transfer capacity of a power system is nowadays possible thanking the huge progress of different, mainly computer technologies embedded in the electric-power systems. However, before reaching this long-time goal, it is necessary to know as much as possible the power transfer characteristics, inter alia, each power line belonging to a certain power distribution network. However, these characteristics of power lines depend, among other things, on the conditions and circumstances existing in the area through which a certain line passes. This practically means that for reaching the goal known under the name smart grid, it is necessary to have on a disposal a methodology that enables determining these characteristics in all practically possible environmental conditions and circumstances.

Besides, many years in the contemporary world the trend of using common corridors (right-of-way) for the planning and construction of new transmission lines and the other type of metal installations is attended that pass through the uninhabited areas. As a consequence of this, electromagnetically coupled two or more transmission lines, as well as transmission lines and some other type of metal installations, are a very common case in the current electric-power engineering practice. In such cases, interactions between different metal installations are inevitable and the relevant characteristics of power lines are not the same as in the case when there are no surrounding metal installations. For solving the problem of determining the relevant characteristics of these power lines have been so far developed and presented in the contemporary scientific and professional literature several mainly computational methods that take into account many relevant factors and parameters. However, in the case of power lines that pass through urban and/or suburban areas, where, as is well known, the surrounding conditions and circumstances are similar to the previously mentioned ones, the situation is quite different. Until recently, no methodology would be enabling the determination of the inductive influence of surrounding metal installations on the relevant characteristic of power lines as well as the inductive influence of one of the power lines on one of these installations [1].

Relevant Characteristics of Power Lines Passing through Urban Areas. DOI: https://doi.org/10.1016/B978-0-323-91136-8.00003-7

The increasing size of modern power distribution networks as well as the higher operating and short-circuit currents of these networks have been matched by overspreading networks of earth return circuits (different pipelines, different cables, and neutrals in low voltage networks, etc.) close to power distribution lines. This circumstance unavoidably results in a mutual inductive coupling of different supply network types (electricity, water, gas, oil, telecommunications, etc.).

The spatial disposition of all installations existing in urban surroundings, including power distribution lines, is determined mainly by the specific disposition of streets belonging to a certain urban agglomeration. By following strict urban rules, all of them are situated side by side on relatively small mutual distances and in a certain depth in the earth under the street surfaces foreseen for walk-sides. Due to such mutual spatial position, all metal installations surrounding a certain power distribution line become encompassed by the fluctuating magnetic field existing along and around that line. This field induces by one current in each of these installations and each of these currents produces an its-own magnetic field that has a feedback effect on the transmission characteristics of that power line. These facts are known practically from the very beginning of the application of Tesla's alternating currents. However, the relevant parameters of these lines have been for a long period, longer than a 100 years, determined by completely ignoring the existence of surrounding metal installations. Designers and later users of distribution lines were forced to work like this because they did not have at their disposal a method that would allow taking into consideration the influence of surrounding metal installations.

The surrounding metal installations grounded at both ends improve transfer characteristics of each power line and have a favorable influence on the grounding problem of supplied substations by reducing part of the ground-fault current returning to the power system through the earth. However, until recently, none of the existing and applicable methods enabled the determining relevant characteristics of power lines passing through urban and/or suburban areas by taking into consideration the favorable influence of surrounding metal installations. Because of that the following question can be posed: Why the problem solution of which would be very desirable and useful was a long time without any usable solution? Certainly, the explanation of this is found in the fact of the practical limitations and impossibilities in collecting numerous relevant data concerning the surrounding metal installations and their spatial positions in urban conditions and circumstances.

The data concerning the structure and topology of a large and complex network, spontaneously formed by all metal installations of any contemporary urban agglomeration, are uncertain or completely unknown. These installations are situated mainly under the surface of the ground, and because of that, many relevant data about them, like their constructive characteristics and their mutual spatial disposition, cannot be visually determined or verified. In some cases, even the total number of metal installations belonging to a certain street is uncertain or completely unknown. Because of these invincible practical obstructions in collecting the necessary input data, planners, designers, and later users of a power distribution network and its lines were former forced to disregard completely the existence of surrounding metal installations when determining relevant characteristics of lines passing through urban and/or suburban areas.

The same practical limitations and impossibilities are the reasons that the determination of the ground-fault current fraction that returns to the power system only through the earth is not possible by measurements either. Thus this fraction of the ground-fault current was a long time unknown, although this fraction is the only one that is relevant for the safety conditions within and in the vicinity of each high voltage (HV) substation. The problem appears because in the case of the substation located in urban surroundings, one more ground-fault current fraction exists and also cannot be determined by the earlier known methods of calculations or by field measurements. This fraction returns to the power system through the grounding system elements (external electrodes) of the supplied substation and metal installations surrounding the feeding line. Both of these two ground-fault current fractions pass through the elements of the grounding system of the supplied substation and continue their flows toward the power system under the surface of the ground. Because of that none of them can be singled out and measured at some accessible and for the measurements suitable place. The process of their mutual separation and dissipation of one of them into the surrounding earth arises spontaneously along many very long external electrodes belonging to the large grounding system of HV substations located in urban surroundings. Because of that the actual value of the reduction factor of an HV or EHV (extra high voltage) distribution line was not possible to determine either by the existing analytical expressions or by measurements.

Besides, there are some more reasons why the previously described problem was considered in scientific and professional circles for a long time as one unsolvable. First of all, this is the fact that the real electric circuit that spontaneously forms along and around any power line that passes through the urban area is very complex. This electrical circuit consists of a large number of mutually different inductively and conductively coupled elements and has a very complex configuration that differs from case to case and cannot be generalized. Therefore even under the unrealistic assumption that all relevant data are known, it is not sure that the problem would be successfully solved in all practically possible conditions and circumstances only by calculations. The existing mathematical models based on the capabilities of modern computers can take into account many relevant factors and parameters but not without a certain less or bigger idealizations and simplifications in relation to the actual physical model.

Because of the high intensity and big asymmetry of the currents flowing through line phase conductors, the effects of the here-considered phenomenon, of inductive interaction between an HV or EHV distribution line and surrounding metal installations, are the most pronounced during ground-fault conditions. At the fault place, ground-fault current leaves phase conductor and returns to the power system by using all available paths, including those that are not foreseen for such function. Due to this, the currents and potentials of high values can appear at places where they normally do not exist and because of that can be potentially dangerous and harmful.

Since power distribution systems consist of single-phase, two-phase, untransposed three-phase overhead and cable lines, as well as cable lines with applied cross-bonding their parameters are not symmetrical. Also, since they serve unbalanced loads, an unbalanced current exists in phase conductors during normal operating conditions, as well. Unbalanced currents in all three-phase conductors are collected in the grounded neutral point of the supplied substation (transformer) and go to the ground through the grounding system of this substation. The return path of this triple current of zero sequences no longer belongs to the electric-power system, so even in normal operating conditions, there is an impact on the environment. Due to significantly lower values of the loud currents, this impact is not so intense, but it is not limited in time as in the case of a ground fault so that the harmful consequences can also be significant, especially when it comes to sensitive electronic equipment. Because of that the considered phenomenon of inductive influence of an HV or EHV distribution line on any of surrounding metal installations deserves to be investigated in normal operating conditions, as well.

The ground-fault current flowing through the earth, the so-called earth current, leaves a power network and returns into it through a large ground volume so that the equivalent depth of the return path is of the order of 1000 m and the corresponding contour of the phase conductor-return path is much larger than in the case of positive (direct) and negative (inverse) system, where currents circulate only through phase conductors. Thus the space in which the electromagnetic field is created by earth current is significantly larger than the space in which the positive or negative electromagnetic field acts. Because of that the inductive/reactive resistance toward the earth current is large. Certainly, this is a favorable fact from the point of view of the intensity of the mentioned dangerous and harmful impacts.

The intensity of the fluctuating magnetic field caused by unbalanced currents is reduced with the currents induced by this field in neutral conductors of power lines. In that way, all the previously mentioned dangerous and harmful effects are reduced, and the achieved degree of reduction is expressed through the coefficient known in the corresponding technical literature as the reduction factor of power lines. Also, metal installations that surround the power lines and are grounded at least at their ends have the same effects as the neutral conductors. Current induced in any of them reduces the part of the unbalanced currents passing through the earth and current induced in each of the neutral conductors and each of remained surrounding metal installations. Besides, they reduce the currents in the phase conductors and in this way improve the transfer capacity of power lines passing through urban and/or suburban areas. Therefore the reaction of the surrounding metal installations on the inductive influence of nearby power lines has several favorable and useful effects on the nearby power lines. Certainly, this is the reason because of which these effects deserve to be determined and utilized in all practically possible conditions and circumstances, especially in urban ones, where they are, certainly, the most pronounced. However, for that, it is necessary to have on a disposal a corresponding methodology.

The analyzes of the complex electrical circuit spontaneously formed in urban and/or suburban areas via an HV or EHV line and surrounding metal installations are enabled by developing the methodology and its versions presented in Refs. [1–9]. The application of this methodology gives possibilities for correct and efficient solving several mutually different and for current power distribution practice important problems. These possibilities can be classified in the following manner:

1. Determination of the actual safety conditions within and in the vicinity of HV substations located in urban areas. This is achieved on the basis of the actual ground-fault current distribution determined by taking into account the inductive influence of metal installations surrounding the feeding line, that is, by determining the actual reduction factor of the feeding line, as can be seen in Chapter 3, Actual reduction factor of power distribution lines.
2. Determination of the actual safety conditions within and in the vicinity of HV substations located in urban areas in special cases where feeding lines are composed of mutually different sections. This is achieved on the basis of the actual reduction factor determined by taking into account all surrounding metal installations and neutral conductors, including those laid only along one section of the whole feeding line, as can be seen in Chapter 4, Reduction factor of lines consisting of one overhead and one cable section, and Chapter 5, Measures for improving reduction factor of a feeding line.
3. Determination of the actual impedance of the grounding system of a supplied substation located in urban areas. This is achieved by eliminating of inductive influence of metal installations surrounding the test circuit on the results obtained by test measurements in the supplied substation, as can be seen in Chapter 6, Preliminary testing of safety conditions of HV substations located in urban areas.
4. Determination of the actual inductive influence of any HV or EHV line on any of the surrounding metal installations during faulty and normal operating conditions. This is achieved on the basis of the actual screening factor determined by taking into account the inductive influence of neutral conductors and all remained surrounding metal installations, as can be seen in Chapter 7, Inductive influence of HV and EHV lines on surrounding metal installations.
5. Determination of actual transfer capacity of HV and EHV lines passing through urban and/or suburban areas. This is achieved on the basis of the actual series impedance of these lines determined by taking into account the favorable inductive influence of surrounding metal installations, as can be seen in Chapter 8, Transfer characteristics of power lines passing through urban areas.
6. Determination of actual ground-fault current at any point along HV or EHV line passing through urban and/or suburban areas. This is achieved by correct determination of actual zero-sequence impedance of these lines by taking into account the inductive influence of the surrounding metal installations, as can be seen in Chapter 8, Transfer characteristics of power lines passing through urban areas.
7. Determination of actual ground-fault current distribution for the fault place at any place along HV or EHV cable line passing through urban and/or suburban areas. This is achieved by determining the actual reduction factor of these lines obtained by taking into account the inductive influence of the surrounding metal installations, as can be seen in Chapter 3, Actual reduction factor of power distribution lines.

All these, from the point of view of the electric-power engineering practice, different practical problems represent in fact only different manifestation forms of one the same physical phenomenon. This is a phenomenon of electromagnetic interaction of inductively coupled electrical circuits formed via conductors an HV or EHV line and the surrounding metal installations with the earth as a common return path.

Before the mentioned methodology, many mutually different methodologies have been developed in the past with the aim to solve at least one of the mentioned problems. These methodologies also enabled the analysis of complex electrical circuits composed of many conductively and inductively coupled elements. However, in the case of power lines passing through urban and/or suburban areas, this is not sufficient. In that case, for solving the quoted problems it was necessary to a methodology that enables the analysis of complex electrical circuits under conditions when many of the relevant data are uncertain or completely unknown.

1.2 The methodology that enables solving the problem

1.2.1 Brief description of the methodology

The development of the methodology presented in this book began with experimental investigations in the 110-kV distribution network of Beograd that have been performed with the aim of determining actual ground-fault current distribution in the grounding systems of the two 110/10-kV substations. These investigations performed in 2006 showed that the previously described problem of deficiency of many relevant data in determining ground-fault current distribution in urban conditions is solvable and the first methodology version that enables this was developed in Ref. [1]. The solution becomes possible thanking the fact that a simulated ground-fault current in any of the phase conductors and the current induced in any of the neutral conductors of tested lines cumulatively involve the inductive influences of all surrounding metal installations. According to this, the developed methodology is based on a ground-fault simulation and measurements of these two test currents. Then, under the assumption that these two currents are in advance known data, the analytical part of the methodology was developed by introducing in the development procedure one fictitious neutral conductor. The problem became finally solvable under the assumption that the imagined physical appearance and spatial position of this conductor are such that it represents a cylinder surrounding all feeding line conductors along the entire line length. The other relevant parameters of this conductor necessary for solving the problem have been determined under the condition that its inductive influence on the ground-fault current distribution is identical with the inductive influence of all, known and unknown, metal installations laid along and around a considered line.

Later on, the described methodology has been modified for the need of solving the problem in the case of overhead power lines. This has been enabled by defining the place along these lines, where the current induced in the neutral conductor (ground wire) should be measured [2]. Thus the methodology version that enables the determination of a current induced in any of the surrounding metal installations, as well as a ground-fault current fraction that, in urban environments, dissipates through the grounding system of the supplied substation into the surrounding ground was rounded. The analytical procedure for the determination of the relevant parameters of the equivalent neutral conductor for any HV or EHV distribution line has been developed. The exception to this was only cable lines with applied cross-bonding and lines composed of one cable and one overhead section. The current that is measured in one of the neutral line conductors in these cases besides induced contains one more component and as such cannot be used as the input data for the methodology at that stage of its development.

As it has been mentioned, the introduced fictitious conductor is cylindrical and it surrounds all power line conductors over the entire line length. In accordance with this, it is clear that this conductor acts as an additional neutral conductor of power lines passing through urban and/or suburban areas and as such improves their transfer characteristics. However, the problem of determining the relevant parameters of this conductor in the case of cable lines with cross-bonding and lines composed with different sections cannot be solved by using the mentioned measurement procedure. To solve the problem in these cases, a new version of the methodology was necessary. This version is based on the use of one of the remaining two phase conductors of the considered line, instead of the previously used neutral conductor when forming the necessary test circuit [7–9]. Thus when all relevant parameters of the introduced fictitious conductor are determined for a certain power line, its actual transfer characteristics can be obtained by treating this conductor as an additional neutral conductor in the well-known calculation procedure for the determination of series impedance of power lines, for example, Ref. [10].

By any of the methodology versions [1–9], a complex electrical circuit with an arbitrary number of conductively and inductively coupled elements and an arbitrary number of unknown relevant data is reduced to a very simple power line model. This line model in ground-fault conditions consists of one phase and one equivalent neutral conductor, whereas in normal operation it contains all phase conductors and one equivalent neutral conductor. Depending on the kind of the problem that should be solved, this equivalent conductor can substitute: all surrounding metal installations or all surrounding metal installations, including neutral conductors of the considered line. Only in the case when it is necessary to determine the inductive influence of a certain power line on one of the surrounding metal installations, this installation that is exposed on only one section of the inducing line is not substituted by an equivalent neutral conductor. The position of the mentioned equivalent conductor, in that case, is imagined in a somewhat different way. It is imagined so that it represents a screen along and around the metal installation for which the inductive influence of a nearby power line should be determined. According to this, in that case this conductor can be called the equivalent screening conductor of the metal installation exposed to the inductive influence of the nearby power line.

The completely developed methodology enables solving all types of practical problems emanating from the physical phenomenon of inductive interaction between power lines and different surrounding metal installations typical for urban areas. All developed methodology versions are based on the same analytical procedure that uses Carson's theory of return path of the alternating current through the earth [11,12], one cylindrical equivalent conductor, as well as the rules well known from the electrical circuit theory.

Finally, all test measurements and calculations necessary for the application of the developed methodology are relatively simple and easily feasible for all types of here-mentioned problems and all types of power distribution lines in all practically possible conditions and circumstances. Thanking all of this, the presented methodology can be without any delay applied in any country over the world, independently of a degree of its technological development and financial possibilities.

1.2.2 Carsons's modified equations and current return path through the ground

Since the distribution feeding line is inherently unbalanced, an unbalanced current leaves the power system through the grounding system of the supplied substation and by passing through the ground returns into the power system through the grounded neutral point(s) in the supply network. Therefore to determine the relevant parameters of a certain power line, it is necessary to know the characteristics of the ground as a conductive medium representing an element of the electrical circuit formed by the power line conductors and the earth as a common return path.

In 1926, Carson managed to develop the equations that enable determining the self and mutual impedances for the arbitrary number of overhead conductors [11,12]. Thanks to these equations for any multiphase power line, we can write an appropriate system of equations and by solving it, with the help of a matrix calculus and modern computers, determine the series impedance of that line.

By developing these equations, Carson assumed that the earth is an infinite, uniform soil with a flat, uniform, upper surface and a constant resistivity. Any end effects introduced at the neutral grounding points are not large at power frequency and are, therefore, neglected.

Carson made the use of conductor images in the way that every conductor at a given distance above ground has its image the same distance below ground surface as is illustrated in Fig. 1.1.

Starting from the general equations of the magnetic field and referring to the conductors presented in Fig. 1.1, the original Carson equations obtained the following mathematical form:

$$Z'_n = R'_n + 4\omega P_{nn} G + j\left(X_n + 2\omega G \cdot \ln \frac{S_{nn}}{RD_n} + 4\omega Q_{nn} G \right), \quad \Omega/\text{mile}, \tag{1.1}$$

$$Z'_{nm} = 4\omega P_{nm} G + j\left(2\omega G \cdot \ln \frac{S_{nm}}{RD_{nm}} + 4\omega Q_{nm} G \right), \quad \Omega/\text{mile}, \tag{1.2}$$

where Z'_{nn} is the self-impedance of conductor n per unit length (Ω/mile); Z'_{nm} is the mutual impedance between conductor n and m per unit length (Ω/mile); R'_{nn} is the resistance of conductor n per unit length (Ω/mile); ω ($\omega = 2\pi f$) is the system angular frequency in radians/second; G is the 0.1609344×10^{-1} Ω/mile; RD_n is the radius of conductor n in feet; GMR_n is the geometric mean radius of conductor n in

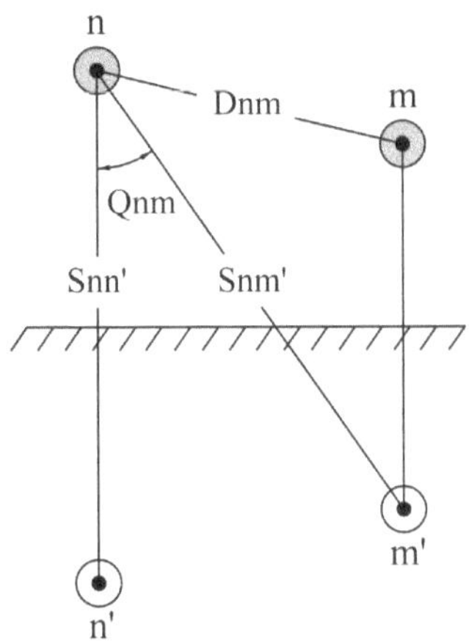

Figure 1.1 Conductors and their images.

feet; f is the power system frequency in hertz (Hz); ρ is the equivalent soil resistivity in Ω-meters (Ωm); D_{nn} is the distance between conductor n and m in feet; S_{nm} is the distance between conductor n and m in feet; and Q_{nm} is the angle between a pair of lines drawn from conductor n to its image and the image of conductor m.

$$X_n = 2\omega G \cdot \ln\frac{RD_n}{GMR_n}, \quad \Omega/\text{mile}, \tag{1.3}$$

$$P_{nm} = \frac{\pi}{8} - \frac{1}{3\sqrt{2}}k_{nm} \cdot \cos(\theta_{nm}) + \frac{k_{nm}^2}{16} \cdot \cos(2\theta_{nm}) \cdot \left(0.6728 + \ln\frac{2}{k_{nm}}\right), \tag{1.4}$$

$$Q_{nm} = -0.0386 + \frac{1}{2}\ln\frac{2}{k_{nm}} + \frac{1}{3\sqrt{2}} \cdot k_{nm} \cdot \cos(\theta_{nm}), \tag{1.5}$$

$$k_{nm} = 8.565 \times 10^{-4} \cdot S_{nm}\sqrt{\frac{f}{\rho}}. \tag{1.6}$$

Somewhat later, by introducing certain simplifying approximations, these equations were modified into a simpler mathematical form. Such modified, these equations are used for the determination of self and mutual impedances of phase and neutral conductors of any type of power line when its constructive characteristics are known, for example, Ref. [13]. In the general case, they have the following form:

$$Z'_n = R'_n + \omega\frac{\mu_0}{8} + j\omega\frac{\mu_0}{2\pi}\ln\frac{\delta}{r_n}, \quad \Omega/\text{km}, \tag{1.7}$$

$$Z'_{nm} = \omega\frac{\mu_0}{8} + j\omega\frac{\mu_0}{2\pi}\ln\frac{\delta_t}{D_{nm}}, \quad \Omega/\text{km}; \tag{1.8}$$

where R'_n is the longitudinal resistance of conductor n, Ω/km; r_n is the mean radius of conductor n in meters (m); D_{nm} is the distance between conductors n and m in meters (m); ω_t is the angular power system frequency: $2\pi f$; μ_0 is the magnetic permeability of vacuum: $4\pi \cdot 10^{-7}$ Vs/Am; μ_r is the relative magnetic permeability; and δ is the equivalent earth penetration depth (m).

The unbalanced current is dissipated through the ground by encompassing a large space around and along the power line. The influence of such a scattered current and the ground as a conductive media on the line impedance in the modified equations has been replaced by the influence a fictitious equivalent conductor placed in the earth at the depth determined by the following expression:

$$\delta = 658\sqrt{\frac{\rho}{f_n}}, \quad \text{m}, \tag{1.9}$$

where ρ is the equivalent soil resistivity around and along the line (Ωm), and f_n is the nominal power system frequency (Hz).

If there were no neutral conductors of line, each unbalanced current would return through the earth and nearby metal installations would be exposed to the influence of an intensified magnetic field. The other extreme is that the specific resistance of the soil is infinitely high, so all this current is returned through neutral line conductors. Then the electromagnetic field generated by the current through the phase conductors and the current through the neutral conductors is closed in a relatively narrow space, so its influence on the surrounding metal installations is negligibly small. In reality, however, an unbalanced current is returned to the power system partly through neutral conductors and partly through the ground. When the lines pass through urban and/or suburban areas, there is another part of this current. That is the part of the ground-fault current that returns into the power system through the metal installations that are laid along the same streets as the considered line. What is the quantitative relationship between the different parts of an unbalanced current in different urban areas is the main question that is solved by developing the methodology that is presented in this book.

For the analysis of power lines in ground faulty conditions, it is important to have in mind something else. The zero-sequence impedance Z_0 differs from positive and negative sequence impedances Z^+ and Z^- because the unbalanced (triple zero sequences) current returns through the ground, while the positive and negative currents circulate through the phase conductors and their sum in all three phases, in each moment and at each place along the line, equals zero. Due to this, the electromagnetic field of ground-fault current captures a much larger space than the space captured by electromagnetic fields created by the action of positive and negative currents. From this, it can be concluded that the energy of the electromagnetic field generated by the ground-fault current is significantly higher than those in the electromagnetic fields created by currents of positive and negative sequence. This is, certainly, a favorable circumstance from the point of view of the impact on the environment, because the line with its large zero-sequence impedance limits unbalanced currents, both those during normal operating conditions and those that occur in conditions of a ground fault. The intensity of this electromagnetic field and the consequent impact of power lines on their environment are reduced by currents induced in neutral conductors and surrounding metal installations, which will be shown in Chapters 2–7. This reduction is of special importance for the power line in normal operation because at the same time it means the reduction of a need for reactive energy necessary for producing fluctuating magnetic field as well as the reduction of losses caused by a transfer of this energy through the power system. This reduced transferring of reactive energy enables by the same extent increased transferring of active energy through the transmission lines supplying distribution networks. Thus the final effect of metal installations surrounding power distribution lines is very similar to the one that is achieved by the well-known compensation of reactive energy, as can be seen in Chapter 8, Transfer characteristics of power lines passing through urban areas.

Here, it is also important to know that the fraction of any unbalanced current returning through the ground follows the route of the power line, which does not have to be the shortest path between the ends of that power line. The reason for this is in a fact that the zero-sequence impedance and its reactive part to be overcome are the lowest when the flow direction of this current coincides with the power line route.

1.2.3 Simplified calculation of the series impedance of HV and EHV lines

Based on the methodology presented in this book, an arbitrary number of neutral conductors and metal installations surrounding a power line can be substituted with only one equivalent neutral conductor. This conductor is in the shape of a cylinder that encompasses all phase conductors on the whole length of the considered power line. Its relevant parameters are the longitudinal electrical resistance and the mean radius of the imaginary cylinder representing this conductor. Since the mean radius, in accordance with the considerations given in Chapter 3, Actual reduction factor of power distribution lines (Theoretical foundations for introducing equivalent sheath) can be freely chosen, and the equivalent neutral conductor can always be determined to represent a common sheath for the phase conductors of the line. When these parameters are determined, by using Carson's modified equations each particular HV or EHV line can be presented with the following primitive matrix:

$$\left[Z_{primitiv}\right] = \begin{matrix} Z_{RR} & Z_{RS} & Z_{RT} & Z_{Req} \\ Z_{SR} & Z_{SS} & Z_{ST} & Z_{Seq} \\ Z_{TR} & Z_{TS} & Z_{TT} & Z_{Teq} \\ Z_{eqR} & Z_{eqS} & Z_{eqtT} & Z_{eq} \end{matrix} \tag{1.10}$$

where Z_{RR}, Z_{SS}, and Z_{TT} are the self-impedance of the phase conductors; Z_{RS} (Z_{RS}), Z_{RT} (Z_{TR}), and Z_{ST} (Z_{TS}) are the mutual impedances between the phase conductors; Z_{eq} is the self-impedance of the introduced equivalent neutral conductor; and Z_{Req}

(Z_{eqR}), Z_{Seq} (Z_{eqS}), and Z_{Teq} (Z_{eqT}) are the mutual impedances between the phase conductors and the equivalent neutral conductor.

Based on this matrix and Kron's reduction technique, for example, Ref. [10], the phase impedance matrix is determined by

$$[Z_{RST}] = [Z_{ij}] - [Z_{in}][Z_{nn}]^{-1} \times [Z_{nj}], \tag{1.11}$$

whereas the sequence impedance matrix is determined by

$$[Z_{012}] = [A]^{-1} \times [Z_{RST}][A], \tag{1.12}$$

where

$$[A] = \begin{bmatrix} 1 & 1 & 1 \\ 1 & -0.5 - j0.866 & -0.5 + j0.866 \\ 1 & -0.5 + j0.866 & -0.5 - j0.866 \end{bmatrix} \tag{1.13}$$

Since the equivalent neutral conductor is imagined as a cylinder around all phase conductors along their entire line length, all mutual impedances between the phase conductors and the equivalent neutral conductor are equal to each other, or

$$Z_{Req} = Z_{eqR} = Z_{Seq} = Z_{eqS} = Z_{Teq} = Z_{eqT}.$$

In the case of single-core cables laid in a trefoil formation, all mutual impedances between the phase conductors are also equal to each other, or $Z_{RS} = Z_{RS} = Z_{RT} = Z_{TR} = Z_{ST} = Z_{TS}$. Thus by using certain algebra manipulations, it is not difficult to show that in that case the phase impedance is determined by the following relatively simple expression:

$$Z'_{RST} = Z'_{ph} - \frac{Z'^2_{pheq}}{Z'_{eq}}, \quad \Omega/\text{km}, \tag{1.14}$$

whereas positive, negative, and zero-sequence impedance of the line, in that case, are determined by the following, also relatively simple expressions:

$$Z^+ = Z^- = Z'_{ph} - Z'_m, \quad \Omega/\text{km}, \tag{1.15}$$

$$Z^0 = Z'_{ph} + 2Z'_m - 3\frac{Z'^2_{pheq}}{Z'_{eq}}, \quad \Omega/\text{km}, \tag{1.16}$$

where Z'_{ph} is the longitudinal self-impedance of the phase conductor; Z'_{pheq} is the longitudinal mutual impedance between the phase conductor and the equivalent neutral conductor, determined by Eq. (1.7); and Z'_m is the longitudinal mutual impedance between two phase conductors, determined by Eq. (1.8).

The given expressions can also be used for the single-core cables laid in a flat formation. In that case, as the mutual distance between the phase conductors should be considered, their mean geometric distance determined by

$$d = \sqrt[3]{d_{12}d_{13}d_{23}}, \quad \text{m}, \tag{1.17}$$

where d_{12}, d_{13}, and d_{23} are the distances between single-core cables laid in flat formation.

According to Eq. (1.10) by introducing an equivalent neutral conductor, the calculations to be performed for lines with an arbitrary number of neutral conductors, including surrounding metal installations are reduced to solving the matrix 4×4.

1.3 Benefits of the methodology application

1.3.1 Grounding problem of HV substations located in urban areas

The results obtained by the experimental investigations and first developed methodology version [1,2] shown that the fraction of the ground-fault current dissipated through the grounding system of the substation supplied by feeding lines passing through the different areas with an average degree of urbanizations is 2.2, 3.2, and 4.5 times smaller in relation to its value obtained without taking into account the surrounding metal installations. This fact throws a completely new light on the grounding problem of HV/medium voltage (MV) substations located in urban or suburban areas and dramatically changes our earlier perception concerning the magnitude of this problem. Now, this problem can be seen within realistic frameworks that, in the general case, enable less conservative solutions that can be in many cases more economical in comparison with those in the past when the complete insight into the ground-fault current distribution was not possible. Greater economical effects certainly can be expected in critical cases where, because of the high soil resistivity and/or high short-circuit current level, special measures (e.g., bare copper wire laid in the same trench as the feeding cable line, counterpoises) are considered necessary. Besides, one can expect the elimination of the strict requirement for the application of expensive MV cables acting as grounding conductors (cables with an uncovered metal sheath) as was the case with the MV cable network of Beograd. The only practical difficulty arises from the fact that the actual ground-fault current distribution depends on many local factors and parameters so that for determining actual ground-fault current distribution each concrete distribution line must be considered individually. Finally, the research results presented in Refs. [2,9] show that disregarding the existence of surrounding metal installations can lead to incorrect testing and wrong definite assessment concerning safety conditions in some critical cases of HV distribution substations. Namely, by using the previously used testing procedure, the required safety criteria in some cases can be considered being satisfied, although too high potential differences and human life losses are possible. This problem is a subject of the separate consideration presented in Chapter 6, Preliminary testing of safety conditions of HV substations located in urban areas.

1.3.2 Problem of the inductive influence of HV and EHV lines

The determination of the inductive influence of power lines on different nearby metal installations had been a dealing subject of many researchers for at least six last decades. During this period hundreds of research papers, as well as many national and international directives (e.g., CCITT, The International Telegraph and Telephone Consultative Committee) were also published to enable determining this influence in different practically possible conditions and circumstances. However, the previously developed methods were applicable only in a small and limited number of practically possible cases, where a scenario of the right-of-way is relatively simple and where all necessary data are available.

On the basis of the physical appearance and spatial position of the introduced equivalent neutral conductor substituting all surrounding metal installations, this conductor acts also as one additional screening conductor for each HV or EHV line passing through urban or/and suburban areas. In this way, it reduces the harmful inductive influences on each of the surrounding metal installations considered individually. The quantitative analysis performed in Ref. [7] shows that in the case of the cable line with applied cross-bonding (the most common cases in the current power distribution practice), the induced voltage in one newly built surrounding metal installation during normal operating conditions is reduced by 70% and 81% in the different urban areas that can be considered average from the standpoint of the degree of urbanization. By the same lines under a ground-fault condition, the induced voltages in one newly built surrounding metal installations are reduced by 54% and by 67%, respectively. These results significantly change our earlier perception concerning the magnitude of this problem. By taking into account the screening effect of the surrounding metal installations, we put this problem into the realistic frameworks and in many critical cases eliminate a need for undertaking the protection measures against too high induced voltages and currents.

The methodology in Ref. [7] had been developed for solving the most complex case that can be encountered in the contemporary power distribution practice. This is the case where the inducing line is an HV or EHV cable line laid in urban and/or suburban areas, where there are many other, known and unknown, metal installations. However, it can also be applied to cases that are solvable thanks to formerly developed methodologies. For example, it can also be applied in the cases where the inducing line is an overhead line constructed through uninhabited areas. These cases are outside of the thematic framework of this book but should be mentioned because of the following reason. The advantage of the methodology presented here in relation to former methodologies is that its application does not require the collection and processing of large amounts of data, as is the case when solving problems requires the use of the capabilities of modern computers.

1.3.3 Influence of surrounding metal installations on transfer capacity of power lines

Based on the physical appearance and spatial position of the introduced equivalent neutral conductor substituting all surrounding metal installations, it is clear that it acts as one additional neutral conductor for power lines passing through urban or/and suburban areas and as such improves the power transfer characteristics of these lines. The results of experimental investigations [1,2], as well as the developed methodology [3,8] enabled the quantitative analyses and insight into the effects of the surrounding metal installations on transfer characteristics of different types of MV, HV, and EHV distribution lines passing through the areas.

Since HV and EHV cable lines with applied cross-bonding are the most common cases in the contemporary power distribution practice, the methodology version applicable in this case is separately developed and considered in Ref. [8]. Based on the performed quantitative analyses, it is shown that the series impedance of a 110-kV line laid in the two different urban areas is 32.85% and 39.02% smaller when we take into account the influence of surrounding metal installations. The final effects are similar to the ones that could be obtained by the well-known compensation of reactive energy. Namely, because of the reduction of the inductive component of the phase impedance (or inductance), the need for the reactive energy in the HV and EHV cable lines is also reduced to the same extent. Due to this, production, as well as the transfer of reactive energy through the transmission network, including these cable lines, necessary for the fluctuating magnetic field along and around the considered cable lines, is also reduced. According to this, it is clear that the surrounding metal installations only by their attendance improve the energy transfer efficiency of the power system having HV and EHV cable lines in its power distribution networks.

The reduction of losses in power transfer occurs due to the very existence of the HV and EHV cable lines so that this improvement on the transfer efficiency appears spontaneously or without our needs and intentions. However, the evaluated favorable effects of surrounding metal installations on the power transfer capacity of power distribution lines can completely be utilized only intentionally. Before that it is necessary to determine the series impedances of these lines by taking into account the inductive influence of the surrounding metal installations. Then, by using these impedances as a basis for power flow studies, it is necessary to determine the actual power transfer capacity of the existing HV and EHV distribution networks that is to some extent greater than was considered before. Certainly, the capacity of active energy transfer is increased by as much as the flow of reactive energy is reduced thanks to the surrounding metal installations. Thus when we take into account the favorable influence of the surrounding metal installations, the more optimum utilization of the existing HV distribution networks by preventing them from being oversized becomes possible. In other words that would enable us to postpone the investments into the new very expensive HV and EHV lines. As it is well known, this is also the reason because which we invest in the realization of the contemporary concept of the more efficient utilization power distribution networks known under the name smart grid. However, in comparison with this conception, the here-presented methodology can be applied immediately and without any investments in the new embedded systems and new equipment, as well as in specially educated personnel for the work and maintaining of these systems and equipment.

Since the HV distribution network represents a very expensive resource in each large town in the contemporary world the favorable economic effects that emanate from the complete utilization of the influence of surrounding metal installations [8], seen on the global level and per 1 year, could be expressed in billions of USD. Since these installations exist independently of our need and desire for more efficient power delivery, their inductive influence on HV and EHV cable lines can be considered a valuable gift for power distribution companies all over the world.

References

[1] Lj.M. Popović, Influence of metal installations surrounding the feeding cable line on the ground fault current distribution in supplied substations, IEEE Transactions on Power Delivery 23 (4) (2008) 2583–2590.

[2] Lj.M. Popović, Testing and evaluating grounding systems for substations located in urban areas, IET Generation, Transmission and Distribution 5 (2) (2011) 231–238.

[3] Lj.M. Popović, Transfer characteristics of electric-power lines passing through urban and suburban areas, International Journal of Electrical Power & Energy Systems 56 (0) (2014) 151–158.

[4] Lj.M. Popović, Determination of actual reduction factor on HV and MV cable lines passing through urban and suburban areas, Facta Universitatis, Series: Electronics and Energetics 2014 (1) (2014) 25–39.

[5] Lj.M. Popović, Ground fault current distribution when ground fault occurs in HV substations located in an urban area, Progress in Electromagnetic Research B 59 (2014) 167–179.

[6] Lj.M. Popović, New method for determination of inductive influence of one or more power lines on any exposed metal installations, International Journal of Engineering Research and Applications 6 (2) (2016) 36–46.

[7] Lj.M. Popović, Inductive influence of HV cable lines in urban and suburban areas, International Journal of Electrical Power & Energy Systems 176 (2019) 105944. Available from: https://doi.org/10.1016/j.epsr.2019.105944.

[8] Lj.M. Popović, Inductive influence of surrounding metal installations on power transfer capacity of HV cable lines with applied cross-bonding, International Journal of Electrical Power & Energy Systems 120C (2020) 105963. Available from: https://doi.org/10.1016/j.ijepes.2020.105963.

[9] Lj.M. Popović, Preliminary testing and assessment of safety conditions of HV substations located in urban areas, Electric Power System Research 195 (2021) 107111. Available from: https://doi.org/10.1016/j.epsr.2021.107111.

[10] W.H. Kersting, Distribution System Modeling and Analysis, first ed., CRC Press LLC, 2002, pp. 95–105.

[11] J.R. Carson, Wave propagation in overhead wires with ground return, Bell Systems Technical Journal 5 (1926) 539–554.

[12] J.R. Carson, Ground return impedance: underground wire with earth return, Bel System Technical Journal 8 (1929) 94–98.
[13] Short-circuits currents in three-phase A.C. system-Part 3: Currents during two separate simultaneous line-to-earth short circuit and partial short-circuit flowing through earth, IEC Int. Std. 60909-3, (Ed.) II, 2003.

Calculated reduction factor of feeding cable lines

2

2.1 Introduction

2.1.1 Definition and significance of the feeding line reduction factor

Conductors in distribution networks are separated from the earth and from each other by a variety of insulation materials: air, paper, or polymers. Occasionally, an unpredictable event ruptures this installation, creating a short circuit between conductors or between conductors and earth. Short-circuit events are system disturbances that generally produce high-magnitude currents. These are different types of short circuits, such as single-phase to ground, double-phase with or without a simultaneous earth fault, and three-phase short circuits, with or without a simultaneous earth fault.

In HV (high voltage) directly grounded distribution networks, the most frequent type of short circuit is single line-to-earth flashovers, mainly due to environmental conditions such as a lightning strike on the line, an equipment failure at the station, pollution of insulation strings, etc. Since the neutral points of these networks are usually directly grounded, the magnitude of the ground-fault current in these networks is very high and in some distribution networks can be higher than 20 kA.

During a ground fault in a network, high currents and raised potentials appear at places where they normally do not exist. Under such conditions an HV power line represents a very complex electrical circuit containing many conductively and inductively coupled elements. At the place of a ground fault, the fault current leaves the phase conductor and returns toward its sources in the power system by using all available paths. Because of that a ground-fault current, for a fault anywhere in an HV power network, is divided into several fractions returning separately to their sources in the power system. Two of them are of special importance for the designer of a future HV substation and its feeding line. One of them is dissipated into the surrounding earth through the grounding system of the supplied substation and returns into the power system through the earth, while the other returns through the neutral conductor(s) of the feeding line. The first one produces all potentials and potential differences (touch and step voltages) relevant for the safety conditions within and in the vicinity of a supplied substation, while the other causes thermal stresses in the neutral conductor(s) of the feeding line(s). Thus a correct estimation of the ground-fault current distribution is necessary for at least two following reasons: for correct designing of the grounding system of a future substation and for the correct selection of the neutral conductor(s) of its feeding line(s). However,

Relevant Characteristics of Power Lines Passing through Urban Areas. DOI: https://doi.org/10.1016/B978-0-323-91136-8.00007-4

before that all available return paths must be separately analyzed and taken into account.

Since the mentioned fractions of the ground-fault current are mainly determined by the constructive characteristics of the foreseen feeding line, a special parameter of the power lines was established in the professional literature and corresponding technical standards. This parameter is called the reduction factor of a feeding line and is defined as the ratio of the part of the ground-fault current returning exclusively through the earth surrounding this line and the total ground-fault current, for example, Ref. [1]. By this definition, it is assumed that grounding impedances at both line ends, or in the supply and supplied substation, are negligibly small.

Under such assumptions used for the definition of the reduction factor, the fraction of the ground-fault current that appears in the neutral conductor(s) is a consequence solely of inductive coupling between the phase conductor carrying the ground-fault current and neutral conductors (ground wires of overhead lines or metallic sheaths in the case of cable lines). Since this component of the fault current returns to the power system solely through the neutral conductor(s), it does not contribute to the creation of any potential on the grounding system of a supplied substation. Therefore it may be said that the reduction factor is a coefficient that expresses only the influence of inductive coupling between the phase conductor carrying ground-fault current and the neutral conductor(s) of a feeding line. This influence is such that it creates in the feeding line neutral conductor(s) a current that reduces the current dissipated through the grounding system of the supplied substation into the surrounding earth.

Based on the given definition of the reduction factor, a part of the ground-fault current dissipated into the earth through the grounding system of the supplied substation is equal to the product of the reduction factor of the feeding line and the total ground-fault current. Thus it can be said that the correctly obtained value of the reduction factor is of primary importance for the correct estimation of safety conditions within and in the vicinity of a supplied substation. If a value of this factor higher than the actual one is used, the result is a too conservative estimate of the achieved safety conditions and possibly excessive (undue) financial expenses for satisfying the strictly prescribed safety criteria. On the other hand, if a value below the actual one is used, this may result in an erroneous conclusion that the safety requirements are fulfilled even though this is not the case. These are the reasons why it is of prime importance for designers of future substations to have at their disposal a calculation procedure to determine the value of the reduction factor of a feeding line as accurately as possible. Also, with the aim of correct selection of the neutral conductor(s), the users of future HV or extrahigh voltage (EHV) lines, mainly power distribution companies, should have at their disposal a calculation procedure for determining the reduction factor of these lines. For example, if the method of estimation of the reduction factor of a cable line is such that the obtained values are significantly above the one sufficient, the power company will order from manufacturer cables that have a larger cross section of metal sheath than the one that is sufficient. This undue increase in metal sheath cross section is certainly a reason for unnecessary increased expenditures for the production of these cables.

Bearing in mind that the lengths of power cables introduced each year into power distribution networks of large towns are enormous and permanently growing, the total amount of these expenditures cannot be undervalued.

It is also important to know that in three-phase medium voltage (MV) cable networks, where normally the cross-bonding (metallic sheaths transposition) is not applicable, each increase of the cross section of cable metal sheaths increases the losses of electric energy caused by circulating currents. Since the circulating currents exist under normal operating conditions, these additional and unjustified losses exist permanently during the whole period of the cable line utilization. Bearing in mind the enormous total lengths of MV cable lines in a large distribution network, it becomes clear that this type of expenditure also deserves full attention. Also, unjustified additional losses in the cable sheaths cause unjustified additional cable heating, which results in a decrease of the ampacity (current-carrying capacity) of these cable lines and their accelerated aging.

The determination of the reduction factor and ground-fault current parts in the ground wire(s) and towers of transmission or overhead distribution lines in a variety of practically possible cases where the line is with uniformly distributed parameters is now a relatively easy task and usually can be solved only by using relatively simple analytical expressions, for example, Ref. [1]. A similar situation is in the case of distribution cable lines, although the considered problem is somewhat more complicated especially in the case where lines consist of three single-core cables. However, the contemporary calculation procedures were preceded by extensive research work that began many years ago and lasted continuously until recently.

2.1.2 A brief review of the existing methods

With the aim of solving the problem of determining ground-fault current distribution, extensive and continuous research work has been carried out over the five decades. The first developed methods for solving this problem are related to the case where the feeding line is constructed as an overhead line with uniformly distributed parameters along its length, for example, Refs. [2–10]. Then, the method based on computer possibilities enables the determination of the critical fault position that yields the maximum earth current through a generating substation grounding system, for example, Refs. [4,5]. Somewhat later the analytical expressions developed in Ref. [8] enabled solving this problem without using modern computers. Subsequently, the papers referring to the problem in special cases, that is, when a feeding line appears as a longitudinal combination of one cable and one overhead section, have been published, for example, Refs. [11–13]. Also, several methods have been developed for solving the problem in critical cases where, because of high local soil resistivity and/or a high level of the short-circuit currents, special measures (e.g., bare copper conductor laid in the same trench as the feeding cable line, and counterpoises) are considered indispensable for achieving the required safety conditions. All of these measures are undertaken with the aim of reducing the part of the ground-fault current dissipating through the grounding system of the supplied substation into the surrounding earth usually named "earth current," for example, Refs. [14,15]. Then, the papers [10,16,17] present the method enabling the determination of the ground-fault current distribution in the case where a feeding line consists of three single-core cables laid in flat or trefoil formations. This method enables taking into account the participation of all three metal sheaths in the ground-fault current distribution for a fault in a supplied substation(s) and at any other place along a cable line consisting of three single-core cables. As such this method served as a basis for the revision of corresponding IEC documents: the International Standard IEC 60909-3 (Ed.) 2 (2003-09) "Short-circuit calculation in three-phase AC system—Part 3: Currents during two separate simultaneous single-phase line-to-earth short circuits and partial short-circuit currents following through earth" and Technical Report IEC 60909-2, Ed. 1 (1992-09) "Short-circuit current calculation in three-phase AC system—Part 2: Data of electrical equipment for short-circuit calculations."

However, the analytical expressions developed for cable lines in papers [10,16,17] correspond to the case of cable lines with no applied cross-bonding that are used nowadays only for very short connections. For any longer HV or EHV cable lines, the cross-bonding technique is required [18]. In the case of these lines, the calculation procedure for determining the reduction factor becomes more complicated and was at the beginning solvable only by using a computer [19]. Then, on the basis of the analytical procedure presented in Ref. [20], it is shown that the reduction factor of these cable lines can also be determined by using relatively simple analytical expressions. It was shown that the mathematical form of these expressions is identical with those relevant for the cable lines with no applied cross-bonding, although the distribution of ground-fault current between the sheaths is changed when we apply cross-bonding.

At the design stage of an HV substation, the knowledge of the value of the reduction factor of its feeding line is necessary for the estimation of the required safety conditions as well as for designing the corresponding grounding system. However, the designers for obtaining data necessary for calculations are forced to make certain assumptions and approximations concerning the data that are not completely available at the design stage. These assumptions and approximations are mainly such that they give final results that are conservative or, in other words, somewhat on the side of increased safety. Such an approach enables finding a rational solution to the problem of grounding, but not also an accurate determination of the reduction factor of the feeding line. Therefore it can be acceptable only if the deviations between the calculated and actual values of the reduction factor are not so great as to lead to large and unnecessary costs in achieving the prescribed safety conditions. However, this is not the case with the feeding lines of HV substations located within urban areas where the inductive influence of surrounding metal installations is not possible to take into account only by calculations. Since in these cases the values of the reduction factor obtained by calculations and its value obtained by test measurements can be essentially different mutually [20,21], we are introducing here two types of reduction factors of these feeding lines. One of them is calculated at the design stage and is named here the calculated reduction factor, whereas the other is named the actual reduction factor and can be determined only for the nearly or already constructed substation and its feeding line, as is shown in Chapter 3, Actual reduction factor of power distribution lines.

On the basis of this short review of the so far published research results concerning the determination of the calculated reduction factor for different types of cable lines, it can be said that all necessary researches are done and that now no one important question or problem has remained without the appropriate answer and solution. According to this, it can be said that the existing knowledge in this field is sufficient for a correct determination of the calculated reduction factor of cable lines in all practically possible cases from the standpoint of the constructive characteristics and spatial disposition of three single-core cables.

2.2 Analytical expressions for the reduction factor

2.2.1 Lines consisting of one single-core or one three-core cable

With the rapid increase in demand for electric energy and the trend for large infrastructures and the vast expansion of highly populated metropolitan areas, the use of underground power cables has grown progressively over the years. Also, in the last decades, three separate single-core cables for distribution cable lines are used instead of three-core cables.

In a single-core distribution cable, normally a metal sheath is coated over the insulation layer to prevent the ingress of moisture and protect the core from possible mechanical damage. Also, it serves as an electrostatic shield (the electric field is enclosed in between the conductor and the sheath) and acts as a return path for fault current and capacitive charging currents.

When an insulated single conductor cable carries alternating current, an alternating magnetic field is created along and around it. If the cable has a metal sheath, it will be encompassed by this field so that the metal sheath of single-core cables acts as a secondary of a large and specific spontaneously formed transformer and the current in the phase conductor induces a voltage in the sheath. Since the sheaths of single-core cables are bonded and grounded at the cable line ends, as common practice for multiconductor cable lines is, the induced voltage causes current to flow in the sheaths, so-called circulating currents during the normal operating conditions.

With the aim to obtain a better insight into the mutual relationship between relevant factors and parameters influencing ground-fault current distribution, let us start our consideration from the simplest case that is rare in practice but the most suitable to analysis. This is the case when substation A supplies substation B through a single-phase feeding line consisting of only one single-core cable. Assuming further that the ground-fault place is in substation B and that the neutral point in this substation is ungrounded, the ground-fault current I_f is distributed as shown in Fig. 2.1.

The notation has the following meaning: U is the voltage source in substation A, I_f is the ground-fault current, I_{sh} is the ground-fault current fraction through the cable sheath, I_e is the ground-fault current fraction returning through the earth, Z_A (Z_B) is

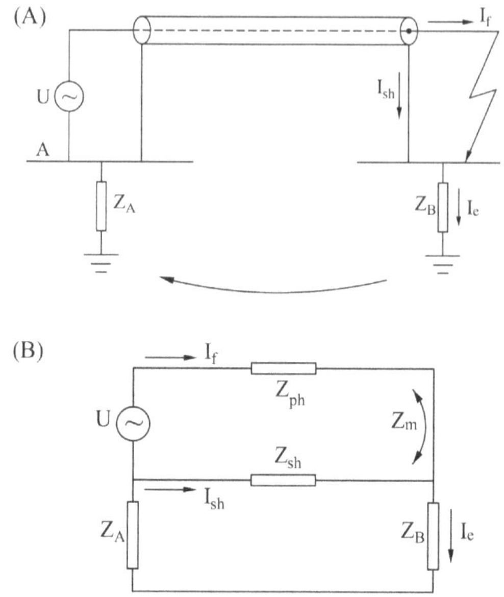

Figure 2.1 The simplest case of the ground-fault current distribution. (A) Physical model of the assumed cable line. (B) Equivalent circuit of the assumed cable line.

the impedance of the grounding system of station A (B), Z_{ph} is the self-impedance of the phase conductor, Z_{sh} is the self-impedance of the cable metal sheath, and Z_m is the mutual impedance between the phase conductor cable metal sheath.

On the basis of the equivalent circuit in Fig. 2.1, one can see that the phase conductor and cable sheath with the earth as the common return path form two mutually coupled electrical circuits. These two circuits are inductively coupled through the impedance Z_m and conductively through the grounding systems of substations A and B. However, under the assumption that the grounding impedances at the line ends are negligible ($Z_A \approx Z_B \approx 0$), the fault current appearing in the cable sheath (neutral conductor) is a consequence solely of the inductive coupling with the phase conductor.

On the basis of the equivalent circuit in Fig. 2.1, one can write the following equation:

$$(Z_s + Z_A + Z_B)I_{sh} + (Z_m + Z_A + Z_B)I_f = 0 \tag{2.1}$$

By using this equation, we can express the sheath current I_{sh} as a function of the total ground-fault current, I_f. Also, for a better insight into the considered problem, we will, by using the decoupled method, represent this current through its two components, for example, Ref. [5]. One of them appears only as a consequence of the inductive coupling between the phase conductor and cable sheath, while the other is a consequence only of the conductive coupling between the cable sheath and the grounding electrodes at the line ends. Thus we can write

$$I_{sh} = I_i + I_c, \tag{2.2}$$

where

$$I_i = -\frac{Z_m}{Z_{sh} + Z_A + Z_B} I_f, \tag{2.3}$$

$$I_c = -\frac{Z_A + Z_B}{Z_{sh} + Z_A + Z_B} I_f. \tag{2.4}$$

According to Eq. (2.3), it is obvious that current I_i represents the component of the fault current appearing in the cable sheath only as a consequence of inductive coupling with the phase conductor (Z_m). The second component of the fault current I_c appears in the cable sheath only as a consequence of conductive couplings with the grounding electrodes of substation *A* and substation *B*.

On the basis of the equivalent circuit in Fig. 2.1 as well as Eqs. (2.2)–(2.4), the total reduction of the ground-fault current through the grounding electrode of substation *B* can be expressed by:

$$r_t = \frac{I_f - I_{sh}}{I_f} = \frac{Z_{sh} - Z_m}{Z_{sh} + Z_A + Z_B}. \tag{2.5}$$

Coefficient r_t involves the effects of both inductive (Z_m) and conductive (Z_A and Z_B) coupling of the cable sheath (Fig. 2.1B). Since the influence of the conductive coupling depends on the local factors like Z_A and Z_B, the coefficient r_t cannot represent the parameter of the cable line itself. To eliminate this influence we will assume, in accordance with the reduction factor definition, that impedances Z_A and Z_B are negligibly small ($Z_A \approx 0$ and $Z_B \approx 0$). Then, by introducing these approximations into Eq. (2.5) and by dividing impedances Z_{sh} and Z_m with the total cable line length, we obtain that the reduction factor of the feeding line consisting of one single-core cable is

$$r = \frac{Z'_{sh} - Z'_m}{Z'_{sh}}, \tag{2.6}$$

where impedances: Z'_{sh} and Z'_m are given per unit length. These impedances are, according to [1], determined by

$$Z'_{sh} = R'_{sh} + \omega\frac{\mu_0}{8} + j\omega\frac{\mu_0}{2\pi}\ln\frac{\delta}{r_{sh}}, \quad \Omega/\text{km} \tag{2.7}$$

$$Z'_m = \omega\frac{\mu_0}{8} + j\frac{\mu_0}{2\pi}\ln\frac{\delta}{r_{sh}}, \quad \Omega/\text{km}, \tag{2.8}$$

where R'_{sh} is the longitudinal cable sheath resistance (Ω/km), r_{sh} is the mean radius of the cable sheath (m), ω angular frequency $\omega = 2\pi f$, μ_0 is the magnetic permeability of vacuum: $4\pi \times 10^{-7}$ (Vs/Am).

The equivalent earth penetration depth is determined by

$$\delta = 658\sqrt{\frac{\rho}{f_n}}, \quad \text{m}, \tag{2.9}$$

where ρ is the equivalent soil resistivity along and around the cable line (Ωm), and f_n is the nominal power frequency (Hz).

Since the cable sheath is relatively thin, the reactance due to the flux inside the sheath, according to Eq. (2.7), is disregarded. Thus by using Eqs. (2.7) and (2.8), we obtain that the reduction factor of the considered cable line is determined by

$$r = \frac{R'_{sh}}{R'_{sh} + \omega(\mu_0/8) + j(\omega\mu_0/2\pi)\ln(\delta/r_{sh})}. \tag{2.10}$$

As can be seen on the basis of Eq. (2.10), reduction factor, r, is not dependent on the cable line length that is also a local factor but only on the constructive characteristics of the used single-core cable.

According to the presented analytical procedure, it is clear that this expression is adequate in the case of cable lines consisting of one three-core cable, as well. However, in the past, it was used for many years for the lines consisting of three single-core cables. Certainly, this practice was wrong and harmful since it without any reason completely disregarded the participation of two other cable line metal sheaths in the ground-fault current distribution.

2.2.2 Relevant characteristics of lines consisting of three single-core cables

Expression (2.19) in the case of these lines is not adequate because they represent, in the general case, an assembly of three single-core cables laid in the ground in different ways, that is, in a trefoil or flat formation. When they are in a flat formation, they may touch each other or may be laid at a certain separation from each other. Due to this the reduction factor of the cable lines consisting of three single-core cables depends not only on their constructive characteristics but also on their spatial formation and mutual distances in each concrete case.

Here, one should also mention in more detail the relevant constructive characteristics of single-core cables. All polyethylene and cross-linked polyethylene (XLPE) cables for voltages above 1.8/3 kV should have a metal screen or a metal sheath. (In the further text only the second term will be used). Its main function is to enclose the electric field inside the cable. This is accomplished by surrounding the insulation of a metal conductor with one or more metal tapes, or braid, or a concentric layer of wires, or a combination of wires and tape(s). A total cross section of the sheath has to be so large that the sheath can withstand thermal stresses, especially those during ground faults. Besides, cables planned for applications where they are exposed to large mechanical stresses have also mechanical protection, that is, metallic armor.

The problem of determination of the reduction factor for cable lines consisting of three single-core cables was for a long time without an adequate solution. Manufacturers of the cables were providing data regarding the reduction factor that refers only to one completely singled out single-core cable on the way in which it is produced and sold. However, distribution companies use them mainly in the formation of three single-core cables laid in the common trench, whereas their metal sheaths are mutually connected and grounded at least at the cable line ends. This obvious misunderstanding between manufacturers and electric-power distribution companies certainly contributed that this problem that remained for a long time hidden, that is, without an adequate solution. One more reason for this could be the fact that the calculated values of the reduction factor of the cable line, in general, cannot be verified by measurements on-site. This is due to the inductive influence of surrounding underground metal installations that are unavoidable in urban and suburban areas. Thus for many years for all cable lines, analytical expression (2.10) was used, although it is correct only for three-core cable lines or the cable lines with only one single-core cable. As a consequence, the values of the calculated reduction factor of many distribution cable lines were considered nearly three times larger than the ones that would be obtained by using the analytical expressions that take into account all three metal sheaths.

For an illustration the ground-fault current distribution in the case of a cable line consisting of three single-core cables is shown in Fig. 2.2.

The used notation has the following meaning: A (B) is the supply (supplied) substation; I_f is the ground-fault current; 1 is the cable sheath of the cable carrying the ground-fault current (in the further consideration faulty cable); 2, 3 are the metal sheaths of the remaining two single-core cables; I_1, I_2, and I_3 are the ground-fault current fractions through the individual cable sheaths; and I_e is the ground-fault current fracture returning into the power system (substation A) through the earth.

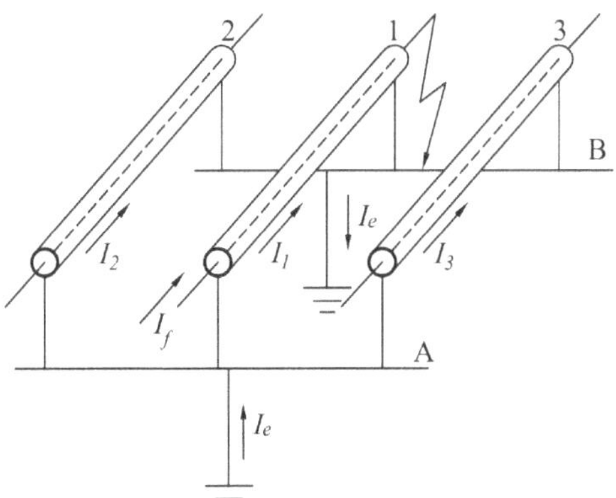

Figure 2.2 Ground-fault current distribution.

Thus it was necessary for the analytical expressions enabling correct determination of the ground-fault current distribution for a fault in the supplied substation, as well as at anywhere along a cable line consisting of three single-core cables. These analytical expressions are derived in Refs. [10,16] and the whole procedure of their deriving is presented here.

2.2.3 Reduction factor of lines consisting of three single-core cables

Let us consider this problem through an example of a cable line belonging to the so-called radial distribution network and whereby generating substation *A* supplies HV distribution substation *B*. The neutral point(s) in substation *A* is (are) either directly grounded or grounded through a low impedance. Thus a ground fault in the feeding cable line or a supplied substation acts as a short circuit and a large current flows through the line phase conductor on the sections between the fault point and grounded neutral point(s). Then, let us assume that the neutral point(s) of the HV distribution network to which the assumed line belongs is (are) grounded only in substation *A*. Under such assumptions the fractions of the ground-fault current in the case of a ground fault in supplied substation *B* are presented by the equivalent circuit shown in Fig. 2.3.

The notation used has the following meaning: *F* is the fault place; I_f is the ground-fault current; I_1 is the fault current through the metal sheath of the faulty cable; I_2 and I_3 are the fault currents through sheaths of the other two cables; I_e is the ground-fault current fraction through the earth; Z_{sh} is the self-impedance of individual sheaths; Z_{01}, Z_{02}, and Z_{03} are the mutual impedances between the faulty phase conductor (denoted by index 0) and individual cable sheaths (denoted by indexes: 1, 2, and 3); Z_{12}, Z_{13}, and Z_{23} are the mutual impedances between individual metal sheaths; Z_A (Z_B) is the impedance of the grounding system of substation *A* (*B*); and *G* is the remote ground.

The balanced components of the ground-fault current circulate only through the line phase conductors and their inductive influence on the nearby parallel circuits is

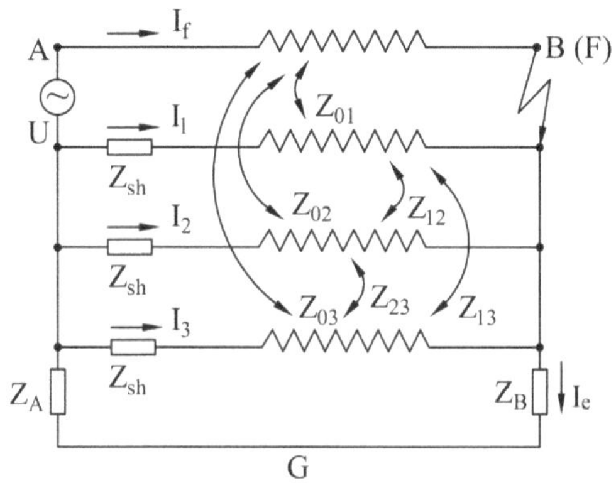

Figure 2.3 Ground-fault current and its fractions.

so small that it can be disregarded. Because of that these currents and the sound phase conductors are disregarded in the presented equivalent circuit.

In accordance with the previously given definition of the reduction factor, in our further consideration, we will assume that impedances Z_A and Z_B are negligibly small ($Z_A \approx 0$ and $Z_B \approx 0$). Also, for further consideration, it is necessary to mention that the current directions shown in Fig. 2.3 are arbitrarily taken.

By using the equivalent circuit in Fig. 2.3 and the definition of the reduction factor, the following system of equations is obtained:

$$
\begin{aligned}
&1.\ Z_{sh}I_1 + Z_{12}I_2 + Z_{13}I_3 + Z_{01}I_f = 0,\\
&2.\ Z_{sh}I_2 + Z_{12}I_1 + Z_{23}I_3 + Z_{02}I_f = 0,\\
&3.\ Z_{sh}I_3 + Z_{13}I_1 + Z_{23}I_2 + Z_{03}I_f = 0,\\
&4.\ r = \frac{I_e}{I_f} = \frac{I_f + I_1 + I_2 + I_3}{I_f}.
\end{aligned}
\tag{2.11}
$$

Since all of the line impedances are proportional to the cable line length, the first three equations in the system of Eq. (2.11) can be divided by the line length. In this manner, one obtains the longitudinal values of these impedances and eliminates the line length as superfluous data. Based on this observation, we can treat all the impedances in Eq. (2.11) as the impedances per unit length and denote them with a prime ($'$). Also, this means that the relative distribution of ground-fault current I_f does not depend on the magnitude of this current. According to this, the only unknown quantities in the given system of equations are I_1, I_2, I_3, and r.

The most general cable disposition, regarding the analytical expressions for mutual impedances: Z'_{02}, Z'_{03}, Z'_{12}, Z'_{13}, and Z'_{23}, is the case when the cables are laid in a flat formation with spacing as shown in Fig. 2.4.

The notation used has the following meaning: d is the distance between two adjacent cables.

Here, it is assumed that the ground-fault current is passing through the core of one of the outer cables as shown in Fig. 2.4. This is the worst/critical case from the standpoint of the safety conditions in a supplied substation because then part of the ground-fault current dissipating through the supplied substation grounding system and the earth takes its largest value. Here, it is interesting to note that this case is also critical from the standpoint of inductive influence on surrounding metal installations. The sum of the currents passing through cable line metal sheaths has its

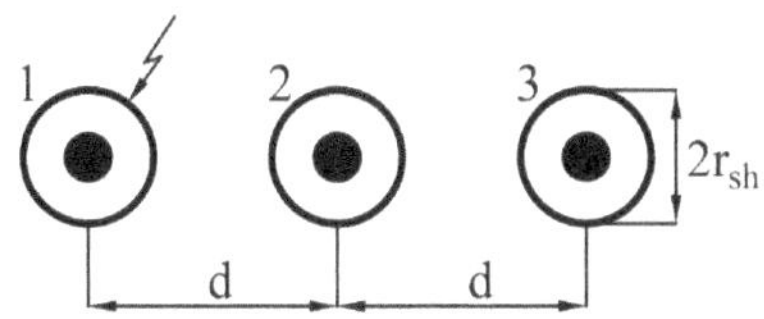

Figure 2.4 Single-core cable flat formation with spacing.

smallest value and because of that, the screening effect of the cable sheaths is also the smallest one.

The mutual impedance between two adjacent cables, denoted in the general case by Z'_{m1}, is determined, according to, for example, Ref. [1], by the following expression:

$$Z'_{m1} = \omega\frac{\mu_0}{8} + j\frac{\mu_0}{2\pi}\ln\frac{\delta}{d}, \quad \Omega/\text{km} \tag{2.12}$$

whereas the mutual impedance between the two outer cables denoted in the general case by Z'_{m2}, is determined by

$$Z'_{m2} = \omega\frac{\mu_0}{8} + j\omega\frac{\mu_0}{2\pi}\ln\frac{\delta_t}{2d}, \quad \Omega/\text{km}. \tag{2.13}$$

Then, on the basis of the given analytical expressions and Fig. 2.4, it is clear that impedances Z'_{02}, Z'_{12}, and Z'_{23} are mutually equal ($Z'_{02} \approx Z'_{12} = Z'_{23}$), and can be jointly denoted by Z'_{m1}. Also, it is clear that impedances Z'_{03} and Z'_{13} are mutually equal ($Z'_{03} \approx Z'_{13}$), and can jointly be denoted by Z'_{m2}.

After introducing these designations in Eq. (2.11) by summing equations: 1, 2, and 3 of this system of equations, one obtains the following analytical expression:

$$r = 1 - \frac{Z'_m + Z_{m1} + Z'_{m2}}{Z'_{sh} + Z'_{m1} + Z'_{m2}}, \tag{2.14}$$

or in a more concise form:

$$r = \frac{R'_{sh}}{R'_{sh} + 3\omega(\mu_0/8) + j3\omega(\mu_0/2\pi)\ln\left(\delta/\sqrt[3]{2r_{sh}d^2}\right).} \tag{2.15}$$

The finally derived expression refers to the general case from the standpoint of spatial formation of single-core cables and to the case when a ground-fault current passes through one of the outer cables. Thus this expression can be used for obtaining the largest earth current I_e, critical from the standpoint of safety conditions within and in the vicinity of the supplied substation.

In the case of cables in a trefoil formation that is frequent in practice, the cables are in a bundle and their positions are such that their centers are placed on top of an equilateral triangle, that is, all three cables are positioned at the same mutual distance: $d \approx 2r_{sh}$. By taking into account this fact instead of Eq. (2.15) one obtains the following expression for the reduction factor of cables laid in a trefoil formation:

$$r = \frac{R'_{sh}}{R'_{sh} + 3\omega(\mu_0/8) + j3\omega(\mu_0/2\pi)\ln(\delta/\sqrt[3]{4}\mathrm{r_{sh}}).} \tag{2.16}$$

On the basis of expressions (2.10) and (2.16), it can be seen that the value of the reduction factor will be larger if the value of the mean radius of the cable sheath, r_{sh}, is larger for identical longitudinal resistance of cable sheath, R'_{sh}. Also, on the basis of Eq. (2.15), it can be seen that the value of the reduction factor will be larger if the mutual distances between the single-core cables are larger for identical R'_{sh} and r_{sh}.

One more effect, very favorable, from the standpoint of grounding problem of the supplied substations, should be mentioned. When the natural conditions for solving grounding problems are less favorable, or when the specific soil resistivity ρ is higher, the value of r is lower, that is, the part of the ground-fault current dissipated into the earth, I_e, is smaller under identical all other relevant conditions. According to expressions (2.10) and (2.16), in the case of lines consisting of three single-core cables, this favorable effect is more pronounced than in the case of lines performed by three-core cables.

On the basis of the presented analytical expressions, it can be noticed that in the case of a flat formation of single-core cables, two different analytical expressions can be derived for the reduction factor of one cable line. One of them refers to the central position of the faulty cable, while the other refers to the other two possible positions of the faulty cable. Certainly, the difference between these two reduction factors is larger if the mutual distances between the single-core cables are greater.

Under normal operating conditions, some currents also circulate in the line metal sheaths grounded at both ends. They appear because the magnetic fields produced by the load currents in each of the phase conductors do not mutually cancel completely, as is the case with the magnetic fields produced by symmetrical phase currents in the metal sheath of the three-core cables. These induced currents, so-called circulating currents, are larger if the spatial arrangement of the single-core cables is less symmetric and if the separations between the cables, because of better cooling, are larger. Certainly, these currents cause permanent losses in the cable sheaths and additional heating in all three single-core cables. Also, as a harmful consequence of these currents, it should be mentioned that this additional cable heating reduces to a certain extent the ampacity of this type of cable line consisting of three single-core cables.

This problem can significantly be mitigated and practically eliminated by applying the technical solution known as the cross-bonding technique. This solution is based on the transposition of the cable sheaths, or the phase conductors together with their metal sheaths, at certain places along the line. The more efficient way for the complete elimination of these currents is that the cable sheaths are grounded at only one point. However, the application of this solution, because of the dangerous and harmful voltages that may be induced in the cable sheaths, is limited on very short lines, mostly those used for local connections within large generating substations.

2.2.4 Reduction factor of cable lines with applied cross-bonding

With the aim to eliminate harmful circulating currents through the sheaths of HV and EHV cable lines, the route of these cable lines is divided into major sections, each of which is divided into three subsections. In each of these subsections, the three sheaths connected in series are associated with conductors of different phases, and their voltages, and consequently the currents through the sheaths have the same magnitude but with a phase shift of 120 degrees. The overall effect is such that the resultant voltage and the current in each of the sheaths are approximately equal to zero during normal operating conditions.

Due to these exchanges in the position of cable sheaths at certain places along the line, the considered problem becomes more complex at first glance, so it was initially solving with the help of a computer [19]. Much later than the beginning of the application of this technique, it was shown that it is possible to derive appropriate analytical expressions for the calculation reduction factor of these cable lines [20].

To obtain the analytical expressions for the reduction factor in this case of HV and EHV cable lines, let us consider ground-fault current distribution in the supplied substation by the equivalent circuit shown in Fig. 2.5.

In this case, as well as in the previously considered one, we also can obtain two different reduction factors for the same cable line. Based on the equivalent circuit in Fig. 2.5, it is obvious the following: if ground-fault current I_f appears in the central single-core cable (denoted 2), currents I_1 and two I_2 will be induced in the cable sheaths instead of currents I_1, I_2, and I_3 induced in the case shown in Fig. 2.5. Thus as well as in the previous case, we will consider the situation when the reduction factor obtains the higher value that is critical from the standpoint of safety conditions in the supplied substation.

Based on the equivalent circuit in Fig. 2.5 and the fact that mutual impedances Z_{02}, Z_{03}, Z_{12}, and Z_{23} are equal each other, as well as that mutual impedance Z_{03} and Z_{13} are equal to each other, we obtain the following system of equations:

$$\begin{aligned}
&1.\ Z'_{sh}I_1 + Z'_{m1}I_2 + Z'_{m1}I_3 + Z'_{m0}I_f = 0\\
&2.\ Z'_{sh}I_2 + Z'_{m1}I_1 + Z'_{m1}I_3 + Z'_{m0}I_f = 0\\
&3.\ Z'_{sh}I_3 + Z'_{m1}I_1 + Z'_{m1}I_2 + Z'_{m0}I_f = 0\\
&4.\ r = \frac{I_e}{I_f} = \frac{I_f + I_1 + I_2 + I_3}{I_f}.
\end{aligned} \tag{2.17}$$

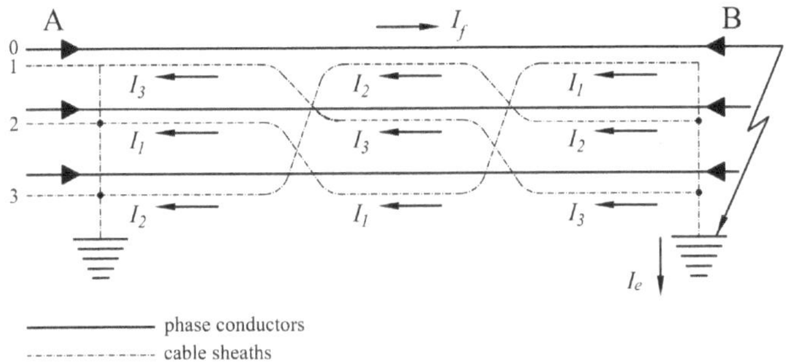

Figure 2.5 Equivalent circuit of the cable line with cross-bonding.

Mutual impedances Z'_{m0} and Z'_{m1} in Eq. (2.17) are determined by the following expressions:

$$Z'_{m0} = \frac{(Z'_{01} + Z'_{12} + Z'_{13})}{3}, \quad \Omega/\text{km}, \tag{2.18}$$

$$Z'_{m1} = \frac{(2Z'_{12} + Z'_{13})}{3}, \quad \Omega/\text{km}, \tag{2.19}$$

whereas the self-impedances of cable sheaths and mutual impedances between the faulty phase conductor and each of the sheaths of a cable line, as well as the mutual impedances between the cable sheaths themselves are given by Eqs. (2.6), (2.7), (2.12), and (2.13).

After summing equations: 1, 2, and 3, we reduce the system of Eq. (2.17) to the system of two equations with unknown quantities I_e and $(I_1 + I_2 + I_3)$. Then, after certain algebraic manipulations, one obtains that the reduction factor in the considered case is also determined by expression (2.15). Also, in the special case, when the cables are laid in a trefoil formation by using the same analytical procedure, one obtains that the reduction factor is determined by Eq. (2.16).

As can be seen, the derived analytical expressions are identical with those relevant to the same cable line, but with no applied cross-bonding. Thus it can be said that cross-bonding does not influence the value of the calculated reduction factor. The only difference stems from the fact that in that case currents I_1, I_2, and I_3 are mutually different, whereas, in the case of applied cross-bonding, they are mutually equal. However, in the cases where the three sections (Fig. 2.5) obtained by cross-bonding are not with the lengths completely equal to each other, the reduction factor is somewhat different for each phase conductor and expression (2.15) or (2.16) in that case are not quite accurate. They can be used only for an approximate estimation under the condition that the mutual differences of the line section lengths are not too large. Also, it is not difficult to see that in the case of a fault anywhere along the line, the distribution of the ground-fault current between the cable sheaths is more balanced in the case of a cable line with applied cross-bonding.

Based on the mathematical form of Eqs. (2.10), (2.15), and (2.16), it is not difficult to see that all three cable line sheaths can be substituted with only one, equivalent sheath, self-impedance of which is

$$Z'_{eq} = \frac{R'_{sh}}{3} + \omega\frac{\mu_0}{8} + j\omega\frac{\mu_0}{2\pi}\ln\frac{\delta}{\sqrt[3]{2r_{sh}d^2}}, \quad \Omega/\text{km}, \tag{2.20}$$

The relevant parameters of this equivalent sheath are equivalent longitudinal resistance equal $R_{sh}/3$ and equivalent mean radius equals $\sqrt[3]{2r_{sh}d^2}$ in the case of cables laid in flat formation and $\sqrt[3]{4}\, r_{sh}$ in the case of cables laid in three-angular formation.

Analytical expression (2.20) is useful for calculations of total grounding system impedance and transferred potentials at the design stage of HV distribution substations, but also does not exist in the current technical standards, for example, Refs. [1,20,22].

A procedure analogous to the presented one can be used to develop analytical expressions for the determination of the reduction factor for the special case of cable lines where they consist of two or more single-core cables per phase. Based on the procedure presented here, it is rather obvious that the value of the reduction factor, in that case, will successively decrease if we successively increase the number of the single-core cables per each of phases.

Under practical conditions the impedance of the substation grounding system is always somewhat larger than zero ($Z_A > 0$ and $Z_B > 0$). Because of that, during the ground faults, raised potentials appear at both ends of the feeding line. These potentials create one more component of the current in the cable sheaths and thus additionally reduce the part of the ground-fault current returning through the earth. This reduction mainly depends on local circumstances as the line length and the values of the grounding impedances at the feeding cable line ends. Consequently, this reduction cannot be treated as a parameter of the cable line itself and can be determined for each concrete case separately. In the case of a sufficiently long line, the influence of the cable sheaths as a conductive connection with the grounding system of substation *A* can be neglected, and that gives the final results that are on the safe side. Also, in the case where substation *A* is transit one this current in the cable sheaths can be taken into account through the value of grounding system impedance of substation *B*. Namely, then cable sheaths together with the grounding system of substation *A* represents an external grounding electrode for substation *B* and its contribution to the grounding system of this substation can be determined by using expression (2.20). All of these facts should be mentioned in the corresponding technical standards, for example, Ref. [1], because concerning the cable lines with cross-bonding cable lines with applied cross-bonding are the most common in contemporary power distribution practice.

Here, also, the following question can be raised: where the induced currents I_1, I_2, and I_3 continue their flow if substation *A* is a transit one and without grounded neutral point(s)? If one cable line supplies both substations (*A* and *B*) and has identical parameters along the whole of its length, according to Kirchhoff's Rules these currents continue their flow toward the supply substation through the grounding connections in substation *A* and the cable sheaths of the rest of the feeding line. Then, in the cases, where we have two or more transition substations, while all other conditions are identical, the induced currents continue their flow through the grounding connections of all other transition substations and the cable sheaths of the rest of the feeding line. However, the situation is different if any of the transit substations is supplied by a line having some other relevant characteristics. For example, when we have an overhead line, one part of the induced currents continue its flow toward its sources through the grounding system of this transition substation and the earth, while the rest of this current continues its flow through the ground wire(s) of the overhead section of the feeding line.

Based on the presented analytical procedure, it is obvious that the reduction factor of a cable line consisted of three single-core cables is not technical data that can be obtained from a manufacturer of these cables. Its value depends, inter alia, on the spatial formation/arrangement in which these cables will be laid. Thus the only designer of a future cable line has all data necessary for the correct determination of its calculated reduction factor. Here, it is also interesting to mention that the reduction factor of an HV and EHV cable line represents always a complex number, but for practical purposes, only its effective value is customarily mentioned.

In the time before the presented analytical expressions were derived, for all cable lines exclusively the expression adequate and correct only for the lines constructed by three-core cables was applying Eq. (2.10). In this manner the influence of the currents induced in the remaining two cable sheaths on the value of the reduction factor of the feeding line is completely disregarded. Certainly, the reduction factor, determined in such a way, was excessively conservative and that led to unnecessarily high consumption of metals—either those used for the cable sheaths or the supplied substation's ground grids. As a final consequence, a significant quantity of useful and expensive metals (copper, aluminum, and lead) in the refined state was needlessly returned into the earth.

2.3 Ground-fault current distribution for any fault place along a cable line

2.3.1 Lines consisting of one single-core cable

In the general case a cable sheath must satisfy some mutually opposed requirements. For example, it has to be with a sufficiently large cross section to withstand thermal stress during ground-fault conditions, but not too large to cause expensive cable production and too high values of harmful circulating currents in the case of the lines with no applied cross-bonding. Therefore it is economically justified to look for as much as a smaller size when selecting the sheath cross section but sufficiently large so that the sheath to withstand thermal stress during ground-fault conditions. To reach this objective, we must have, inter alias, a method, as accurate as possible, for the calculation of the ground-fault current distribution for a fault anywhere along the cable line.

Let us assume a simple case when substation A supplies substation B through a single-phase cable line and that only station A has grounded neutral point(s). Under such an assumption and under an assumption that the fault appears at an arbitrary point, F, along this line, the ground-fault current distribution can be presented as shown in Fig. 2.6.

The notation used has the following meaning:

I_{ai}—current induced in the cable sheath, and I_{ac} (I_{bc})—current appearing as a consequence of conductive connection with the grounding electrode in substation A (B).

According to the adopted assumptions, the total ground-fault current I_f comes only from station A. Thus its inductive influence exists only on the cable section from substation A to fault place, F. If we, according to the decoupled method, for

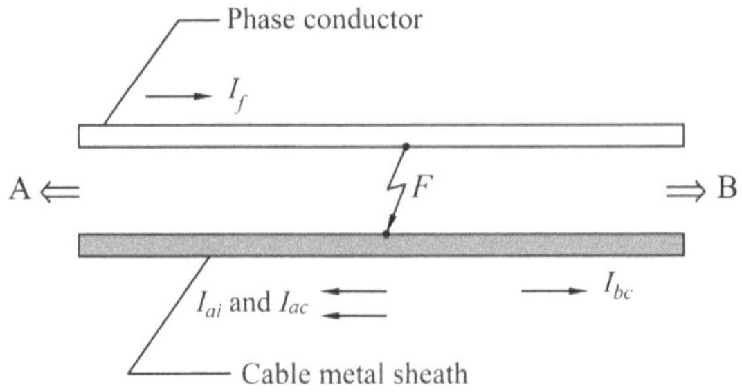

Figure 2.6 A fault at an arbitrary place along the assumed line.

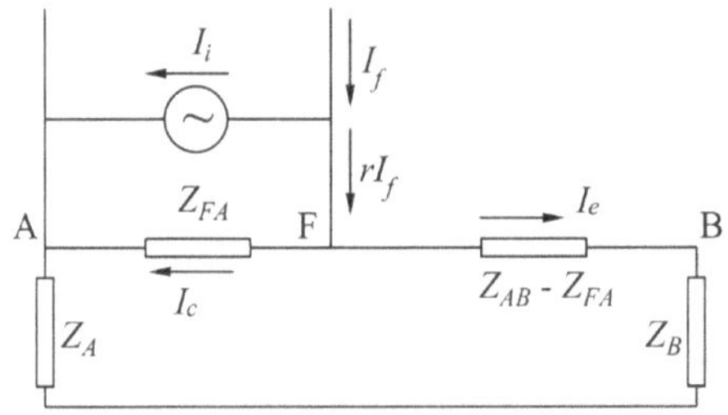

Figure 2.7 The equivalent circuit for a fault at any point along the assumed line.

example, Ref. [5], present this influence separately by one equivalent current source, the ground-fault current distribution can be presented by the equivalent circuit shown in Fig. 2.7.

The notation has the following meaning:

Z_{AB}—the self-impedance of the cable sheath for the whole line, Z_{FA}—the self-impedance of the cable sheath from the substation $A-F$, and r—the reduction factor of the line for the fault at the end of the line.

We can represent impedance Z_{FA} as a function of the distance of the fault place ($l=\overline{AF}$) by using the following expression:

$$Z_{FA} = lZ'_{sh}, \tag{2.21}$$

where l is the cable line length from station A to F (km), and Z'_{sh} is the self-impedance of the cable sheath, per unit length (Ω/km).

Based on the equivalent circuit in Fig. 2.7 the part of the fault current rI_f is divided into two fractions, the currents I_c and I_e. The current I_c directly flows toward substation A, while current I_e flows also toward substation A but via the grounding system of substation B. Thus according to the presented equivalent circuit and under the assumptions: $Z_A \approx 0$ and $Z_B \approx 0$ that is in accordance with the definition of a feeding line reduction factor, we obtain

$$I_e(L-l)Z'_{sh} = (rI_f - I_e)lZ'_{sh}, \tag{2.22}$$

where L is the total cable line length (km).

Then, based on Eq. (2.22) the reduction factor for a fault at any place along the line is determined by

$$r(l) = \frac{I_e}{I_f} = \frac{l}{L} r. \qquad (2.23)$$

where $r\ (l)$ is the reduction factor for a fault at an arbitrary distance l along the line, seen from substation A.

Based on the given expression, it can be seen that the reduction factor linearly increases with the increase of the distance of the fault place seen from substation A. For the fault near the beginning of the cable line, its value is close to zero, and for the fault in substation B, it becomes equal to the reduction factor determined by Eq. (2.10). Here, it is important to note that for the faults near the beginning of the cable line, almost the total ground-fault current returns into the power system through the cable sheath. Because of that the value of the ground-fault current at this place must be considered critical from the standpoint of the thermal stress that a cable sheath has to withstand during the ground fault.

Now, let us consider the simplest case, from the standpoint of ground-fault current distribution, in a three-phase cable line. This is, certainly, the case, where a line is constructed by one three-core cable. Since these cables have only a single metal sheath, it is clear that the derived analytical expressions (2.10) and (2.23) correspond to this type of cable line, as well. Here, it is important to note that inductive coupling between the sheath and the phase conductor(s) depends mainly on the constructive characteristics of the cable themselves and slightly on the local geological characteristics, that is, equivalent soil resistivity, ρ. Thus in the case of this type of power cable line, the reduction factor represents mainly a constructive characteristic of the cable itself and can be required from the cable manufacturer as one of the data relevant for its application in a certain distribution network.

2.3.2 Lines consisting of three single-core cables

Let us assume that a fault appears at an arbitrary place along a feeding cable line consisting of three single-core cables. For simplicity, we will assume that the ground-fault current comes only from substation A and that, by this, the induced component of the fault current in the metal sheath of the faulty cable appears only in the line section between substation A and the fault place. Also, for simplicity and a better insight into the problem, let us neglect a relatively small current that can, in some practical cases, be injected into the ground at the fault place due to the damaged insulation covering the cable metal sheath. Under such assumptions the ground-fault current distribution in the faulted cable only and at the fault place is presented by the equivalent circuit shown in Fig. 2.6.

Based only on the equivalent circuit in Fig. 2.6, it is not difficult to imagine that in the case of a line having three single-core cables, the situation regarding

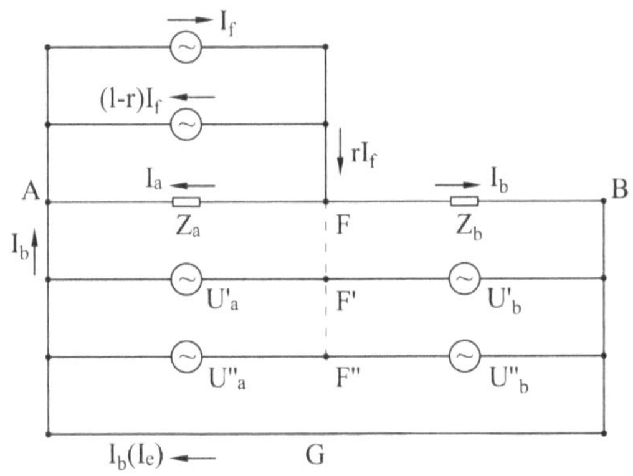

Figure 2.8 The equivalent circuit for a fault at any point of the cable line.

ground-fault current distribution becomes much more complicated. To obtain a better insight into the considered problem, let us apply the decoupled method. In this way, we can present separately the components of the fault current appearing as a consequence of the conductive connections with the grounding electrodes at the line ends. Also, in this way, we can present separately all components of the fault current appearing in the metal sheaths as a consequence of the inductive coupling with the faulty phase conductor. Finally, based on the previous assumptions and the decoupled method, the cable line, for a fault at an arbitrary distance from the supply substation along the feeding line ($l=\overline{AF}$), can be presented by the equivalent circuit shown in Fig. 2.8.

The notation used has the following meaning: U'_a (U'_b) is the voltage source in the metal sheath of the nearer sound cable as a consequence of inductive coupling between this metal sheath and the faulty cable from A to F (from F to B); U''_a (U''_b) is the voltage source in the metal sheath of the more distant sound cable as a consequence of inductive coupling between this metal sheath and the faulty cable from A to F (from F to B); $(1-r)I_f$ is the current source representing the fault current appearing in the metal sheaths as a consequence of their inductive coupling with the faulted phase conductor; I_a and I_b are the left and right current returning into the power system through the metal sheath and appearing as a consequence of its conductive connections; Z_a (Z_b) is the self-impedance of the metal sheath from F to A (B); and F' and F'' are the points on the two sound cables nearest to the fault place, F.

On the basis of the given equivalent circuit, the voltages induced in sections l and $(L-l)$ of the metal sheaths of the sound cables are mutually equal and determined by the following equations:

$$U'_a = U'_b = \frac{l(L-l)}{L} Z'_m r I_f, \tag{2.24}$$

$$U''_a = U''_b = \frac{l(L-l)}{L} Z'_{m2} r I_f. \tag{2.25}$$

Since, according to the given equivalent circuit, these voltages act in mutually opposite directions, they completely cancel each other. Thus regarding the currents in the presented equivalent circuit, we may consider that these voltage sources do not exist. According to this, it is obvious that current I_e is equal to current I_b.

Then, for the remaining part of the circuit shown in Fig. 2.8, we can write the following system of equations:

$$\begin{aligned} &1.\ I_a + I_b = r(l)I_f, \\ &2.\ \frac{I_a}{I_b} = \frac{L-l}{l}. \end{aligned} \tag{2.26}$$

Finally, based on this system of equations and according to the definition of the reduction factor, one obtains the following simple formula for determining ground-fault current distribution for a fault place anywhere along the cable line:

$$r(l) = \frac{I_e}{I_f} = \frac{I_b}{I_f} = \frac{l}{L}r, \tag{2.27}$$

As can be seen, expression (2.27) gives the relationship between the reduction factor when the fault place is at any place along the line and the reduction factor, when the fault is at the line end, that is, in the supplied substation. It shows that the reduction factor for the fault anywhere along the line is a linear function of the line length from the supply substation to the fault place. This observation represents the theoretical foundation for some conclusions important for the contemporary electrical engineering practice. They can be defined in the following manner:

- from the standpoint of safety conditions in the supplied substation, the critical faults are those within the substation itself;
- from the standpoint of inductive influence upon exposed metal installations, the critical faults are also those within the supplied substation.

Then, since expressions (2.23) and (2.27) have the same mathematical form, one more important conclusion should be mentioned. This is the following:

- The function $r(l)$ depends on the number of neutral conductors only through the value of the reduction factor for a fault in the supplied substation, r.

This means that $r(l)$ involves the influence of surrounding metal installations if r is determined by using the procedure that is presented in Chapter 3, Actual reduction factor of power distribution lines. This fact is quite obvious only on the basis of the equivalent circuit presenting the ground-fault current distribution presented, Fig. 2.8.

Also, based on Fig. 2.8 and an analogous analytical procedure, one can show that Eq. (2.27) is relevant in the case of cable lines with applied cross-bonding, as well. According to this and the previous conclusion that cross-bonding does not influence the value of reduction factor r, one can draw the following conclusion:

- The reduction factors of the cable line with no applied cross-bonding and the same cable but with applied cross-bonding have mutually equal reduction factors for the same fault place along each of them, from its beginning to its end.

In other words, although the currents through the individual cable line sheaths are changed, their sum at any fault place along the line remains unchanged after applying cross-bonding.

Finally, based on expression (2.27) it is not difficult to see that in the case of a ground fault near the supply station (at the beginning of the line), the reduction factor $r(l)$ has a value close to zero. This means that practically the whole ground-fault current I_f returns into the power system solely through the cable sheath of the faulty cable. Because of that this current and this fault place must be considered critical from the standpoint of the thermal stress of the cable sheath under the assumption that the ground-fault shutdown time is equal for the fault at any place along the line.

2.3.3 Supplied substations with the grounded neutral point(s)

In practice, especially in the case of HV distribution networks, the neutral points(s) in the supplied substation is(are) also directly grounded, and because of that the ground-fault current distribution along an HV or EHV cable line is more complicated. Namely, the total ground-fault current represents the sum of two components, one arriving from station A (I_{fa}) and one arriving from station B (I_{fb}). In that case, according to the previous consideration and the principle of superposition, the total ground-fault current is distributed in the manner shown in Fig. 2.9.

The notation used has the following meaning: L is the total cable line length; l_a (l_b) is the cable line section between the fault place and substation A (B); I_{fa} (I_{fb}) is the left (right) component of the ground-fault current I_f; I_a (I_b) is the left (right)

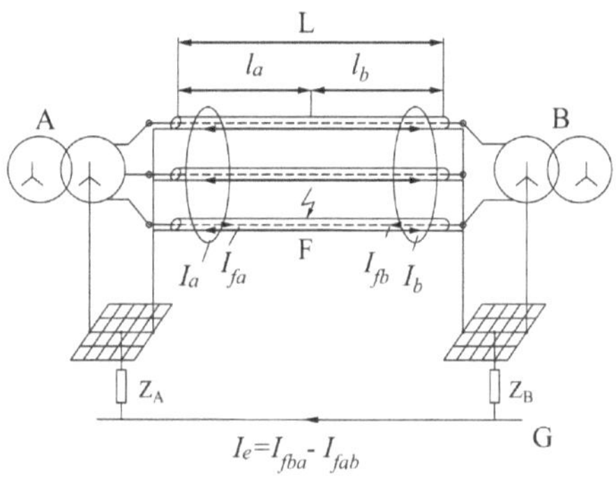

Figure 2.9 Ground fault at an arbitrary place along an HV or EHV cable line. *EHV*, Extrahigh voltage; *HV*, high voltage.

fault current circulating through the metal sheath of the faulty cable; and I_{fba} (I_{fab}) is the current as a consequence of the conductive connection between the fault place F and the grounding electrode of substation A (B).

The ground-fault current distribution in Figs. 2.6 and 2.7 is obtained under the assumption that the insulating sheath covering the metallic sheath at the fault location is not damaged. However, even in that case, practical conditions (very large grounding resistance of cable sheath at the fault place) are such that the current that would inject into the ground, as a consequence of such damage, would be negligibly small.

When the neutral point(s) in substation B is(are) grounded, one component of the fault current comes from this substation. For the faults within substation B itself, this component circulates only in the axial direction through the grounding system elements and cannot create any raised potential on that grounding system of this substation. Because of that this ground-fault current component independently of its value cannot be relevant from the standpoint of the safety conditions in the supplied substation. For the fault at any place along the cable, one part of this current returns to substation B via the sheath of the faulted cable, grounding system of substation A, and the earth as well as the grounding system of substation B, Fig. 2.9. Thus it reduces to some extent the value of the current dissipating through the grounding system of substation B and returns to substation A (earth current I_e), that is, the reduction factor for a fault any place on the feeding cable line. However, this fact has no practical importance, because, as we have already shown, the critical fault position, for the safety conditions within and in the vicinity of a supplied substation, is always this substation itself.

A similar consideration can be performed for a more general and more complicated case when substation B belongs to an HV network with one or more grounded neutral points. In that case the ground-fault current for the fault at any place along the cable line also consists of two components, one arriving from substation A and the other arriving from substation B, but the component from substation B consists of two mutually different subcomponents. One of them is passing through the grounded neutral point(s) in substation B, while the other is returning into the power system through all of the other grounded neutral points in the HV network at the side of substation B. Certainly, the same case is with the component coming from supply substation A which usually belongs to a large transmission network.

Starting from the assumption that the influence of raised potentials in stations A and B is negligible ($Z_A \approx 0$, $Z_B \approx 0$), currents through the metal sheath of the faulted cable, according to Fig. 2.8, are determined by

$$I_a = \left(1 - r\frac{l_a}{L}\right) I_{fa} + r\frac{l_a}{L} I_{fa} \frac{Z_F}{Z'_{sh} l_a} + r\frac{l_b}{L} I_{fb} \frac{Z_F}{Z'_{sh} l_b}. \tag{2.28}$$

$$I_b = \left(1 - r\frac{l_b}{L}\right) I_{fb} + r\frac{l_b}{L} I_{fb} \frac{Z_F}{Z'_{sh} l_b} + r\frac{l_a}{L} I_{fa} \frac{Z_F}{Z'_{sh} l_a}. \tag{2.29}$$

where Z_F is the grounding impedance of the metal sheath at the fault place, F.

This impedance is determined, according to the circuit shown in Fig. 2.7, by

$$Z_F = \left(\frac{1}{Z'_{sh} l_a} + \frac{1}{Z'_{sh} l_b} \right)^{-1}. \tag{2.30}$$

The impedance Z_F represents practically a parallel connection of the two metal sheaths belonging to different line sections, one from the fault place to substation A and the other from the fault place to substation B.

As can be seen, currents through the sheath of the faulty cable, I_a and I_b, consist of one induced and two components that are a consequence of the conductive connections between the place of the ground fault and the grounding electrodes at the line ends.

The currents induced in the sheath of the faulted cable are determined by

$$I_{ai} = \left(1 - r\frac{l_a}{L}\right) I_{fa}, \tag{2.31}$$

$$I_{bi} = \left(1 - r\frac{l_b}{L}\right) I_{fb}. \tag{2.32}$$

Currents I_a and I_b have their maximum values if the faults are very near substation A ($l_a \rightarrow 0$) and substation B ($l_b \rightarrow 0$), respectively. By using the circuit in Fig. 2.9, it is not difficult to show that values of these currents at two critical locations (near station A and near station B) can be evaluated by the following approximations:

$$I_a(l_a \rightarrow 0) \approx I_{fa} + rI_{fb}, \tag{2.33}$$

$$I_b(l_b \rightarrow 0) \approx I_{fb} + rI_{fa}. \tag{2.34}$$

The larger of the currents obtained by these equations is to be applied for the correct selection of the metal sheath capable of withstanding thermal stress during a ground fault. As the impedances Z_A and Z_B are neglected, these equations give results that are slightly on the side of enlarged safety. According to this, expressions (2.33) and (2.34) are necessary for correct selecting the metal sheath of cable used for a certain cable line, which is nowadays regulated mainly through the corresponding national standards.

The presented analytical expressions enable the determination of the ground-fault current distribution for a fault anywhere along the line composed of three single-core cables. This enables us to adopt MV, HV, and EHV cable with a metal sheath that is in accordance with actual needs, that is, large enough to withstand thermal stress in ground-fault conditions, but not too large to cause expensive cable production and too high circulating currents in the case of cable lines with no applied cross-bonding.

2.4 Quantitative analysis

2.4.1 Reduction factor calculated values

By using the presented analytical expressions, let us consider the values of the calculated reduction factor for a 110-kV cable line consisting of three single-core cables in a network with a nominal frequency equal to 50 Hz. The relevant data on single-core XLPE-insulated cables with a copper sheath (copper sheath of wires helically applied) are $r_{sh} = 0.0455$ m and $R'_{sh} = 0.18$ Ω/km (the sheath cross section is 95 mm^2). The cables are laid in trefoil formation ($d = 0.091$ m), whereas the equivalent soil resistivity is $\rho = 100$ Ωm.

At first, let us consider the values of the reduction factor obtained by using Eq. (2.16) and data provided by the cable manufacturer for these cables before publishing the appropriate analytical expressions [10,16]. Also, let us consider the influence of the equivalent soil resistivity along and around the cable line in a wide range of actually possible values, as well as the influence of different cable formations on the value of the reduction factor.

The results of the calculations of the reduction factor effective values are summarized and presented in Tables 2.1–2.3.

The values of $|r|$ obtained by taking into account only one metal sheath or provided by the cable manufacturer are almost three times larger, as shown in Table 2.2.

As can be seen, in all cases presented in Table 2.3, values of the reduction factor increase as the distance between the single-core cables increases and soil resistivity decreases.

Now, let us assume an HV substation that is at the design stage and that will be supplied by a 110-kV cable line with the required effective value of the reduction

Table 2.1 Triangular formation by using Eq. (2.16).

ρ (Ωm)	30	100	300
$\|r\|$	0.1057	0.0992	0.0940

Table 2.2 Data provided by the manufacturer.

ρ (Ωm)	30	100	300
$\|r\|$	0.2763	0.2618	0.2494

Table 2.3 Value of $|r|$ for different flat formations by using Eq. (2.15).

	ρ (Ωm)		
d (m)	30	100	300
0.091	0.1040	0.0988	0.0931
0.25	0.1114	0.1051	0.0951
0.50	0.1186	0.1107	0.1031

factor less than 0.16. The cable type has been foreseen to be XHE49-A. In the standard construction, it has a sheath with an outer diameter of $d = 91$ mm that is made of copper strings with a total cross section of 3×95 mm^2. According to the manufacturer's catalog, the reduction factor for this cross section of the cable sheath is 0.297. By selecting the next larger cross section offered in the catalog, 3×120 mm^2, the reduction factor, also according to the catalog, would be 0.235. Finally, the sheath cross section of 3×185 mm^2 provides, also according to the catalog, the required values of the reduction factor equal to 0.157. Since the length of a feeding 110-kV cable line per one 110/10 kV substation is approximately 4 km (an average value according to the data from the distribution network of Beograd, Serbia), the quantity of the copper, necessary to decrease the reduction factor to the required value, below 0.16, is 9600 kg. However, according to the here-presented analytical expressions, the correct value of the reduction factor of the considered line consisting of the single-core cables with the standard design of metal sheath (Cu 3×95 mm^2) has a value that is significantly lower, that is, about 0.1. Here, it should be mentioned that copper strings with a total cross section of 95 mm^2 withstand fault currents of maximally 21.5 kA for fault duration limited to 0.5 seconds. Thus based on the presented analytical expressions, the application of the larger cable sheath cross section offered by the manufacturer, from the standpoint of the required value of the reduction factor, is nothing else but a redundant expenditure.

Finally, the current electric-power industry standards usually suggest a unique and semiempirical value for the reduction factor for the different MV cable lines. This value is mainly significantly larger than the one that is obtained by Eq. (2.10) or Eq. (2.16), but the safety conditions of the supplied MV substations, because of urban surroundings are always satisfied with a certain reserve on the safe side because the ground-fault currents in MV distribution networks are always limited, usually on the values of 300 or 1000 A. Also, sheath cross sections of these cables are smaller (usually Cu 3×25 mm^2) and their calculated reduction factor obtained by the analytical expressions presented here is significantly higher in comparison to the calculated reduction factor of cable lines in HV networks. However, the total length of the cables in a 10-kV network belonging to one HV/MV kV substation is much larger (of the order of 120 km) so that the total savings obtained by using the here-presented analytical expressions for selecting appropriate cable sheath in MV cables networks could be significant for power distribution companies.

2.4.2 Fault place critical for the metal sheaths of cable lines

To obtain a more complete insight into the distribution of ground-fault current for different fault points along the line, calculations were performed for the currents through cable line sheaths. The results of these calculations are presented graphically and the obtained diagram, representing relative values of the currents induced in the considered cable sheath by current $i_A(I_A/I_{fA})$ and $i_B(I_B/I_{fB})$ as a function of relative distances from station A (l_a/L) in the case of cables laid in trefoil formation and the equivalent soil resistivity equal to 100 Ωm, is presented in Fig. 2.10.

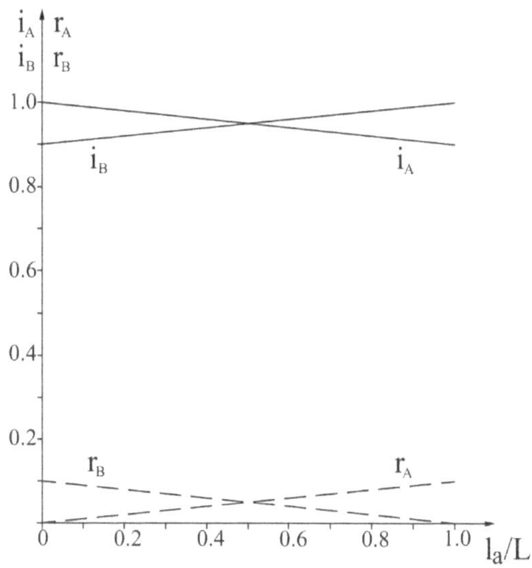

Figure 2.10 Relative values of the currents in the sheath of a faulted cable and the corresponding reduction factors.

As can be seen, the diagrams with full lines representing i_A and i_B are symmetrical with respect to the middle of the cable lines. The diagrams with the dashed lines, representing the reduction factors r_A, and r_B for the right and left ground-fault current components, are also symmetrical in relation to the middle of cable lines.

In accordance with the given diagram, it is not difficult to conclude that current I_A flowing toward substation A reaches its maximum when the fault place is approaching the beginning of the line, while current I_B flowing toward substation B reaches its maximum when the fault place is approaching substation B. Here it should be mentioned that the absolute value of current I_{fA} hyperbolically decreases with increasing the fault distance from substation A, while current I_{fB} hyperbolically decreases with increasing fault distance from substation B. Also, in the case that substation A is transit one, it is interesting to note the following: in modern distribution networks, it is not uncommon for HV stations to change roles at least for a short time and so that the transit substation becomes a supplied substation and vice versa. In that case, as can be seen from the diagram in Fig. 2.10, the relative distribution of ground-fault current for fault anywhere along the cable line remains unchanged that is, linear function of the distance of the fault place from the supply or supplied substation.

2.5 General observations

Single-core cables have been utilized in distribution networks of medium (10–35 kV) and high (66–132 kV) voltages all over the world for many decades.

However, correct determination of the calculation reduction factor of lines made of these cables has become possible only after deriving here-presented analytical expressions and their introduction into international standard [1]. After that it has become possible in solving many practical problems to take into account the participation of all three metal sheaths in the ground-fault current distribution for faults anywhere along power lines consisting of three single-core cables.

Before the presented analytical expressions, the estimation of the safety conditions within and in the vicinity of the HV distribution substations at the design stage was excessively conservative. The estimated values of all potential and potential differences on the grounding systems of all HV substations supplied by cable lines were around 2.5 times larger than those that are estimated by using the here-presented analytical expressions. Depending on other relevant factors such as the level of ground-fault currents, the duration of cleaning time (ground-fault duration), and local soil resistivity in each concrete case, this was the reason for less or more unnecessarily additional consumption of useful metals used for the cable sheaths and/or for the grounding electrodes of supplied substations. A significant amount of useful metals, copper, aluminum, and lead, were returned in the refined state into the earth without an actual need.

Also, in three-phase MV networks, where the cross-bonding is not applicable for the elimination of the circulating currents, each unnecessary increase in the cross section of the metallic sheaths increases energy losses caused by these currents. Since the circulating currents appear under normal operating conditions, these increased losses grow progressively from year to year (each year numerous new MV cable lines are introduced into distribution networks all over the world). Also, these additional losses cause additional heating of the cables, which results in a certain reduction of the ampacity (current-carrying capacity) of the cable lines and of the whole distribution network. From this point of view, critical places are certainly in the vicinity of the HV/MV substations, where a large number of MV cables are concentrated in a narrow space, that is, in a small number of joint trenches.

Since the use of single-core cables is increasing, especially cables with XLPE, it is realistic to assume that a significant saving in the production of this type of cable is accomplished by a wide application of the here-presented analytical expressions. Potential savings are growing with an increase of the total length of the applied cables, that is, with the development of MV and HV distribution networks, as well as with a permanently growing need for the replacement of aged cables.

Bearing in mind all these facts, IEC-TC 73 (Technical Committee for "Short-circuit currents") accepted the proposal of Serbia and Montenegro for revision of the corresponding IEC documents: International Standard IEC 60909-3 (Ed.) 2 (2003-09) and Technical Report IEC 60909-2, Ed.1 (1992-09). By treating the presented analytical expressions as a new technology, IEC-TC 73 (at the meeting in Dresden, September 2004) decided to revise the standard IEC 60909-3 (2003) just 1 year after the beginning of its implementation. Thus the presented analytical expressions served as a foundation for the current version of the mentioned technical standard, IEC 60909-3, Edition 3.0, 2009-02, "Short-circuit currents in three-phase AC systems-Part 3: Currents during two separate simultaneous line-to-earth short circuits and partial short-circuit currents flowing through earth" [1]. Through this technical standard the presented analytical expressions became available for wide application in current electrical engineering practice. However, in the meantime, it has been shown that the presented analytical expressions are identical to those that are relevant to the calculation reduction factor of the lines with cross-bonding [20], so it would be useful in this regard to make an appropriate amendment to this standard.

All types of power cable lines, almost without any exception, are being laid in urban and suburban areas. In such an environment, in addition to metal sheaths of cable lines, many other metal installations participate in returning the ground-fault current toward its sources in the power system. Because of that each HV or EHV cable line forms a very complex electrical circuit with many conductively and inductively coupled elements having uncertain or completely unknown constructive characteristics and spatial position. Thus further investigations in this field were necessary and the obtained results are shown in Chapter 3, Actual reduction factor of power distribution lines.

References

[1] Short-circuits currents in three-phase A.C. systems-Part 3: currents during two separate simultaneous line-to-earth short circuit and partial short-circuit flowing through earth, IEC Int. Std. 60909-33 (0) (n.d.) 2009–03.

[2] J. Endrenyi, Analysis of transmission tower potentials during ground faults, IEEE Transactions on Power Apparatus and Systems 86 (10) (1967) 1274–1283.

[3] R. Varma, D. Mukhedkar, Ground fault current distribution in substations, towers, and ground wires, IEEE Transactions on Power Apparatus and Systems 98 (3) (1979) 724–730.

[4] S.A.P. Meliopoulos, R. Web, E. Joy, S. Patel, Computation of maximum earth current in substation switchyards, IEEE Transactions on Power Delivery 102 (9) (1983) 3131–3139.

[5] S.T. Sobral, V.G.P. Fleury, J.R. Villalba, D. Mukhedkar, "Decoupled method" for studying large interconnected grounding systems using microcomputers—Part I—Fundamentals, IEEE Transactions on Power Delivery 3 (4) (1988) 1536–1544.

[6] B. Tapar, S. Madan, Currents for design grounding systems, IEEE Transactions on Power Delivery 5 (3) (1984) 2633–2636.

[7] Lj.M. Popović, Dissipating of the partial ground fault current across shield wires of transmission lines, Electric Power Systems Research 11 (1) (1986) 25–37.

[8] Lj.M. Popović, Practical method for evaluating ground fault current distribution in station, towers, and ground wire, IEEE Transactions on Power Delivery 13 (1) (1998) 123–129.

[9] Lj.M. Popović, A practical method for evaluation of ground fault current distribution on double circuit parallel lines, IEEE Transactions on Power Delivery 15 (1) (2000) 108–113.

[10] L.M. Popović, Determination of the reduction factor for feeding cable lines consisting of three single-core cables, IEEE Transactions on Power Delivery 18 (3) (2003) 736–743.

[11] S.T. Sobral, et al., Grounding potential rise characteristics of urban step-down substations fed by power cables—A practical example, IEEE Transactions on Power Delivery 3 (2) (1988) 1564–1572.
[12] Lj.M. Popović, Practical method for evaluating ground fault current distribution in station supplied by an unhomogeneous line, IEEE Transactions on Power Delivery 12 (2) (1997) 722–727.
[13] S. Mangione, A simple method for evaluating ground-fault current transfer at the transient station of a combined overhead cable line, IEEE Transactions on Power Delivery 23 (3) (2008) 1413–1418.
[14] Lj.M. Popović, Efficient reduction of fault current through the grounding grid of substation supplied by cable line, IEEE Transactions on Power Delivery 15 (2) (2000) 556–561.
[15] A. Campuccia, G. Zizzo, A study of the use bare buried conductors in an extended interconnected earthing system inside an MV network, in: Proceedings of the CIRED Conference, Turin, 2005.
[16] Lj.M. Popović, Improved analytical expressions for the determination of the reduction factor of the feeding line consisting of tree single-core cables, European Transactions on Electrical Power 18 (2) (2008) 190–203.
[17] Lj.M. Popović, Ground fault current distribution when a ground fault occurs anywhere along the line consisting of three single-core cables, in: Proceedings of the Sixth Symposium, Nikola Tesla, Beograd, Serbia, October 2006.
[18] IEEE Guide for Bonding Shields and Sheaths of Single-Conductor Power Cables Rated 5 kV through 500 kV, IEEE Standard 575-2014, 2014.
[19] R. Benato, et al., Ground return behavior in high voltage alternating current insulated cables, Energies 7 (12) (2014) 8116–8131.
[20] Lj.M. Popović, Preliminary testing and assessment of safety conditions of HV substations located in urban areas, Electric Power System Research 195 (2021) 107111. Available from: https://doi.org/10.1016/j.epsr.2021.107111.
[21] Lj.M. Popović, Influence of the metallic installations surrounding the feeding cable line on the ground fault current distribution in supplied substations, IEEE Transactions on Power Delivery 23 (4) (2008) 2583–2590.
[22] IEEE Guide for Safety in AC Substation Grounding, IEEE Std. 80-2000.

Actual reduction factor of power distribution lines

3

3.1 Introduction

In the case of a substation located in an urban or suburban area, the correct determination of the ground-fault current distribution is not possible on the basis of the analytical expressions for the determination of the cable line reduction factor presented in Chapter 2, Calculated reduction factor of feeding cable lines. All of these analytical expressions have been derived under the assumption that other metal installations in the vicinity of these lines do not exist. However, in practice, all types of power cable lines, almost without any exception, are being laid in urban and/or suburban areas, where many other metal installations exist and participate in returning the ground-fault current into its sources in the power system. Because of that the actual reduction factor of cable lines is not possible to know if we do not take into consideration the influence of the surrounding metal installations involved with the intensive fluctuating magnetic field appearing along and around a high-voltage (HV) or extra-high voltage (EHV) line during a ground fault in the supplied substation.

The permanently increasing sizes of modern urban distribution networks, causing also higher operating and short-circuit currents, have been matched by overspreading networks of earth return paths (different pipelines, neutral conductors of other electric-power lines, etc.) situated in close vicinity of electric power distribution lines. The spatial positions of these installations in each contemporary city have been primarily determined by the mutual disposition, as well as length and width of individual streets. Due to this, all of these installations are independently of their basic function concentrated on the relatively narrow space and the small mutual distances. This unavoidably results in the mutual inductive coupling of each of them with all other of them. Besides, all of these installations are also conductively mutually connected and grounded at all reinforced concrete blocks of foundations of almost all newly built buildings in contemporary towns.

Thus different metal installations such as sheaths of different types of cable lines, neutral conductors of the low-voltage (LV) networks, and steel water pipes are, because of strict rules posed by modern urbanization, situated in a relatively narrow space defined by the width of the streets and their sidewalks. Some of them are only at certain places in contact with the earth, whereas some others are in effective and continuous contact with the earth. Also, they are firmly mechanically interconnected by the application of the terra-neutral (TN) grounding system in consumer installations of all buildings arranged along each street. Thus their spatial dispositions along the streets unavoidably lead to the mutual electromagnetic

Relevant Characteristics of Power Lines Passing through Urban Areas. DOI: https://doi.org/10.1016/B978-0-323-91136-8.00002-5

interaction of various supply networks (electricity, water, gas, oil, telecommunication, etc.). The intensity of this interaction in practice obviously is a very complex function of many different factors and parameters that are different in each concrete case. However, if we wish to determine the actual reduction factor of a certain HV or EHV cable line, we have to determine the inductive influence of surrounding metal installations on the ground-fault current distribution passing through this line.

The grounding system of a distribution HV/medium-voltage (MV) substation consists of the substation grounding electrode and many outgoing MV cable lines acting as external grounding electrodes and/or conductive connections with the grounding systems of the supplied consumption (MV/LV) substations. Thus the grounding system spontaneously formed through outgoing MV cable lines covers a large urban area around an HV/MV distribution substation. Such a grounding system includes, through the TN grounding system in the LV network and consumer installations, many, known and unknown, metal installations typical for an urban area. As a result, the metal sheaths of all outgoing MV cable lines spontaneously become conductive connections with different metal installations situated along the same street(s) as the feeding line of an HV substation. Depending on their spatial arrangement, these surrounding metal installations are more or less inductively coupled with the feeding line of each such HV substation. They spontaneously participate in returning a ground-fault current into the power system and reduces the fraction of this current dissipated through the supplied substation grounding system into the surrounding earth. Since this effect is very favorable, it is useful to take it into account and obtain the actual reduction factor of the feeding line at the time of verifying the safety conditions in a newly built substation before its putting into operation.

The electrical circuit spontaneously formed by metal installations surrounding a certain HV or EHV cable line involves all inductively coupled metal installations having the shape and property of long electrical conductors and situated around and along a considered line. Thus it is not difficult to see that the actual physical model of the spontaneously formed electrical circuit is very complex and that involves numerous different metal installations over a large urban area through which the considered line passes. In conditions when the ground-fault current flows through the phase conductor of an HV or EHV cable line, the currents appear in the surrounding metal installations due to their inductive couplings with this conductor. However, the magnitude of these currents is not dependent only on the inductive coupling with the faulty phase conductor and the value of the ground-fault current passing through this conductor. These currents also depend on the constructive characteristics of the metal sheaths of the considered cable line, the number of surrounding metal installations, their constructive characteristics, their mutual spatial positions, as well as many other factors and parameters of smaller importance. In each of these installations, this current depends, inter alia, on its spatial position toward the faulty phase conductor of the considered cable line and toward all other surrounding metal installations. However, most of these installations are laid under the surface of street sidewalks, and many relevant data about them cannot be visually or in some other way determined or verified. Thus the main problem emanates from the fact that many relevant data about the surrounding metal installations are uncertain or completely unknown. The same or similar can be said for the composition of the soil and its specific resistivity around and along any power line passing through urban and/or suburban areas.

Each of the surrounding metal installations, together with the earth as the common return path, forms one electrical circuit, while all of them together form one complex electrical circuit with a usually unknown number of conductively and inductively coupled elements. Thus the considered problem can be defined in the following manner: how to determine the current induced in these elements? Since for each of these currents, on the basis of Kirchhoff's rules, one can write the corresponding system of equations with currents as unknown quantities, and since the analytical expressions necessary for the self- and mutual impedances of different types of underground conductors are known, it can be said that the previously described problem has been, in principle, solvable long time ago, Chapter 1, Introductory considerations. However, because of objective limitations and impossibilities in collecting data relevant for determining the self- and mutual impedances of surrounding metal installations, the problem of the determination of the actual reduction factor of lines passing through urban areas was a long time in the past considered as unsolvable one.

According to the previous consideration, the problem can also be defined in the following manner: how it is possible to overcome the deficiency of a huge number of relevant data and obtain the actual reduction factor for each particular power line? Thus at the first sight, it seems that for determining the currents induced in surrounding metal installations, the measurements on-site are sufficient. However, whether such measurements are feasible? They are not feasible because the spontaneously formed grounding system of an HV distribution substation, besides the grounding function, acts also as a conductive connection with the metal installations surrounding its feeding HV line, and because of that, during a ground fault, two of the ground-fault current fractions pass through the elements of this grounding system. One of them circulates only in the axial direction through the grounding system elements and continues its flow to the power system through all metal installations involved by the magnetic field created by the ground-fault current in the feeding line. As such this current does not participate in the creation of any potential on the grounding system of the supplied substation. However, neither of these two fractions of the earth fault current can be singled out and determined by direct or indirect measurements.

As an illustration of a ground-fault current distribution for a fault in the supplied substation, the main ground-fault current fractions, including those mutually separated by passing through the grounding system elements, I_i and I_e, are in the general case presented in Fig. 3.1.

The used notation has the following meaning: $A(B)$ is supply (supplied) HV substation; F is ground-fault place (supplied substation); I_f is ground-fault current; I_n is ground-fault current fraction induced in the line neutral conductor(s); I_i is ground-fault current fraction induced in the metal installations surrounding the feeding HV line; and I_e is current that dissipates into the earth through the grounding system of the supplied substation, B.

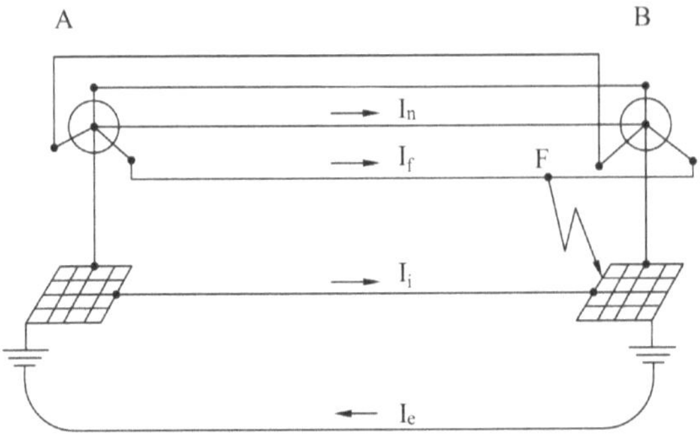

Figure 3.1 Main fractions of a ground-fault current in urban conditions.

According to the equivalent circuit presented in Fig. 3.1, both of the mentioned current fractions, I_i and I_e, flow through the elements of the grounding system of supplied substation *B*. At this, current I_e is dissipated into the surrounding earth, while current I_i continues its flow through the metal installations located along and around the feeding HV line. In the case of a ground fault, these currents after passing through the elements of the grounding system of substation *B* leave the power system and separately continue their flows toward the earthed neutral points in the supply HV network. Since both of these currents leave a power system, all potentially dangerous and harmful influences of an HV distribution substation and its feeding line upon the environment originate from these two ground-fault current fractions. Earth current, I_e, creates all potentially dangerous potentials and potential differences within and in the vicinity of the supplied substation, while induced current, I_i, also can be dangerous and harmful because it flows through the metal installations (different metal pipelines, the metal screen of telecommunication lines, etc.) that are not foreseen for this function. According to this, we have to undertake adequate protective measures against the undesirable effects of these currents. However, this is not possible if we do not know to determine the values of these two currents.

Current I_i cannot be measured because the surrounding installations have no sufficient places accessible and suitable for direct and synchronous measurements of the currents induced in all of them. Such places are neither the metal sheaths of all outgoing MV cable lines because these sheaths represent at the same time connections with the grounding electrodes of many supplied MV/LV substations and the connections with metal installations surrounding the feeding cable line. Thus some parts of earth current I_e also flow through these sheaths and cannot be separated and directly measured or indirectly determined by measuring the potential of the grounding system of substation *B*. As it is well known, in urban conditions, because of the attendance of many other underground metal installations, the correct measurements of this potential are not possible. For the same reason, it is not possible to determine the actual impedance of the grounding system of the supplied substation, which will be discussed in more detail in Chapter 6, Preliminary testing of safety conditions of HV substations located in urban areas.

Finally, a practical impossibility in the estimation of the influence of surrounding metal installations also stems to some extent from the fact that one of the relevant data, the value of equivalent soil resistivity of the surrounding area, cannot be determined exactly. Although there are several methods of measuring soil resistivity, no one of them is applicable in urban conditions. The reason for this comes from the fact that surfaces of urban areas are already covered by buildings, streets, pavements, and many other permanently constructed objects, while under the ground surface many metal installations are already laid.

All previously considered limitations and impossibilities served as the reasons for undertaking the corresponding experimental investigations in the power distribution network of Beograd [1]. As final results of these investigations, somewhat later it was developed the whole methodology that enables the determination of actual reduction factor, actual transfer capacity, and actual inductive influence of the HV or EHV lines that are passing through urban and/or suburban areas [1−12].

3.2 Cable lines with no applied cross-bonding

3.2.1 Complete equivalent circuit

The beginning of solving the previously described problems can be considered experimental investigations performed in 2006 in the 110-kV distribution network of Belgrade [1]. They showed that in the conditions of a simulated ground fault, the currents that occur in the cable line sheaths are significantly less than expected, obtained using the analytical expression (2.16). To examine this, at the first sight, paradoxical effect, during the measurements, a consecutive increase in the number of sheaths grounded at both ends of the line was performed, starting with only one to all three of them grounded at both ends of the line. The measurement results showed that these successive inclusions lead to a reduction not only of the potential at the grounding electrode of the supplied substation but also a reduction of the current/currents through the already included sheath/sheaths. This fact shows that the surrounding metal installations all together act as one additional neutral conductor (cable sheath) and accordingly, all of them can be represented by only one fictitious conductor parallel to the considered cable line. Certainly, the further aim in solving the considered problem was how to determine the relevant parameters of that fictitious conductor? However, before that, it was necessary to obtain a complete insight into the complex electrical circuit spontaneously formed during a ground fault. In the other words, it was necessary an equivalent circuit in which each of the surrounding metal installations is presented separately.

Now, let us assume an HV cable line consisting of three single-core cables, with all three sheaths grounded at both line ends and with no applied cross-bonding. Also let us assume the general case, from the standpoint of spatial formation of the single-core cables, as shown in Fig. 3.2.

The used notation has the following meaning: d is the distance between two adjacent cables, and r_{sh} is the mean radius of the cable sheaths.

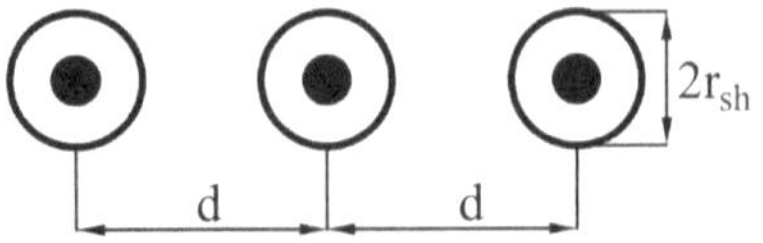

Figure 3.2 Single-core cables in a flat formation.

On the basis of the previous considerations, it is certain that each HV or EHV line passing through urban and/or suburban areas during a ground fault represents a more complex electrical circuit than it can be concluded only based on the total number of its conductors. The ground-fault current circulating through one of the phase conductors induces in each of the surrounding metal installations one electromotive force. Thus in this way spontaneously formed electrical circuits can be imagined, in a general case, as a complex electrical circuit with many different mutually conductively and inductively coupled electrical lines, and with the earth as a common return path. To obtain a necessary insight into this, in many cases in a great deal hidden, electrical circuit, let us assume, for the beginning, that we know each of its constituent elements. Then the total number of conductors to be individually presented is equal to the total number of surrounding metal installations increased by the number of neutral conductors of the line and by one of the phase conductors that is adopted for ground-fault simulation.

As an example, let us consider an HV cable line consisting of three single-core cables, with all three metal sheaths grounded at both line ends. Then, let us assume that the total number of surrounding metal installations, including the cable sheaths, is equal to an arbitrarily large number N that, seen theoretically, can be infinitely large. Finally, the complete equivalent circuit of the assumed line and its surrounding metal installations, under the conditions of a simulated ground fault in the supplied substation, can be presented as shown in Fig. 3.3.

The used notation has the following meaning:

U_a is auxiliary voltage source; U_{n0}, U_{n1}, U_{n2}, ..., U_{nN} are voltages induced in an arbitrary (nth) circuit (metal installation) by the current in each of the surrounding circuits; I_t is simulated ground-fault current through one of the phase conductors of the considered line; I_e are currents injected into the earth through the grounding system of substation B; I_1 is current through the sheath of the cable carrying current I_t; I_2, I_3 are currents through the metal sheaths of the other two single-core cables; I_4, I_5, ..., I_n, ..., I_N are currents induced in the individual surrounding metal installations; Z_0 is the self-impedance of the phase conductor of the considered line; Z_1, Z_2, Z_3 are the self-impedances of the cable metal sheaths; Z_4, Z_5, Z_6, ..., Z_N are the self-impedances of the individual surrounding metal installations; and N is the total number of surrounding metal installations enlarged by the number of the cable sheaths.

Concerning the symbols used in Fig. 3.3, it is also necessary to say that the current directions shown in this circuit are taken arbitrarily.

As can be seen, the presented equivalent circuit is composed of the self- and mutual impedances of all surrounding metal installations, all cable line screens, and phase conductor carrying current I_t, including all induced voltages. The

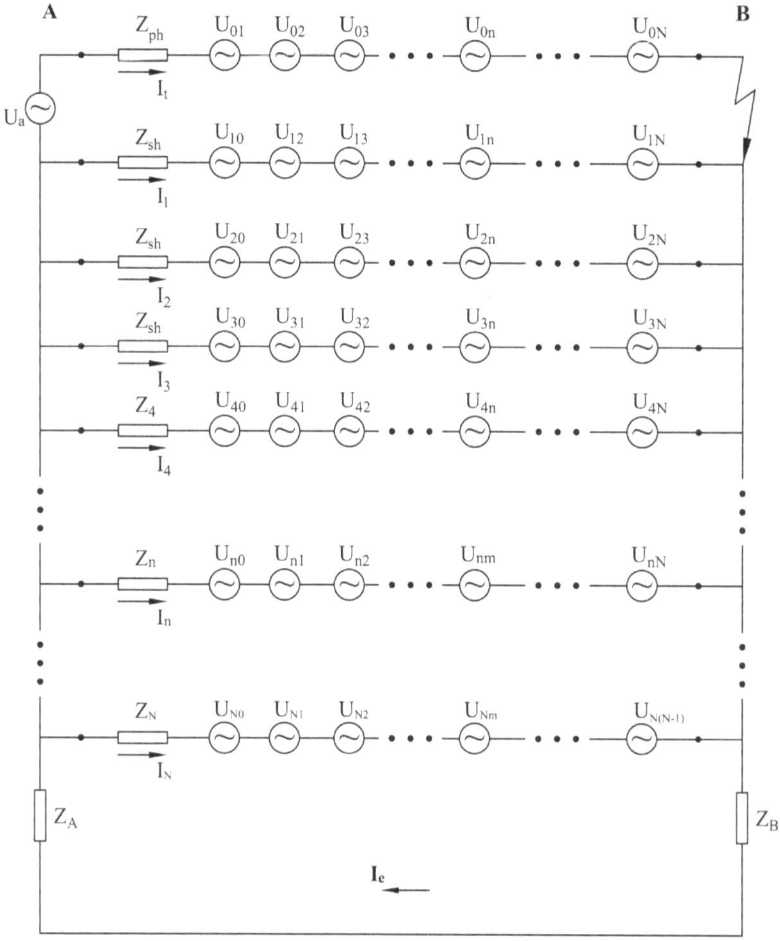

Figure 3.3 Complete equivalent circuit of the assumed line.

self-impedance of an arbitrary, *n*th, surrounding metal installation is, according to for example, Ref. [13], determined by

$$Z'_n = R'_n + \omega_t \frac{\mu_0}{8} + j\omega_t \frac{\mu_0}{2\pi} \ln \frac{\delta_t}{r_n}, \quad \Omega/\text{km}, \tag{3.1}$$

whereas the mutual impedance between two arbitrary, *n*th and *m*th, surrounding metal installations is determined by

$$Z'_{nm} = \omega_t \frac{\mu_0}{8} + j\omega_t \frac{\mu_0}{2\pi} \ln \frac{\delta_t}{D_{nm}}, \quad \Omega/\text{km}; \quad m \neq n \tag{3.2}$$

where R'_n is the longitudinal resistance of an arbitrary, nth, surrounding metal installation, Ω/km; r_n is the radius of an arbitrary, nth, surrounding metal installation (m); D_{nm} is the distance between two arbitrary, nth and mth, surrounding metal installations (m); ω_t is the angular test frequency: $2\pi f_t$; μ_0 is the magnetic permeability of vacuum: $4\pi \times 10^{-7}$ Vs/Am; μ_r is the relative magnetic permeability of the installation material; and δ_t is the equivalent earth penetration depth (m).

The equivalent earth penetration depth is determined by

$$\delta_t = 658\sqrt{\frac{\rho}{f_t}}, \tag{3.3}$$

where ρ is the equivalent soil resistivity along and around the cable line (Ωm); and f_t is the test circuit frequency.

Current in an arbitrary surrounding installation, I_n (Fig. 3.3), induces in an also arbitrary (mth) surrounding installation a voltage, U_{mn}, which is determined by

$$U_{mn} = -Z'_{nm}I_n, \quad \text{V/km}, \quad m \neq n \tag{3.4}$$

where Z'_{nm} is the mutual impedance between the mth and nth surrounding metal installation.

Based on the equivalent circuit shown in Fig. 3.3, it is possible to write the system of N independent equations. Based on it and the known value of the anticipated total ground-fault current (available on the basis of separate study) and values of all self- and mutual impedances, it is possible to determine each of the induced currents. However, it is sufficient that only one of the self- or mutual impedances in the presented equivalent circuit cannot be determined and the whole problem becomes unsolvable. Unfortunately, because of the previously mentioned practical limitations and impossibilities, numerous data necessary for determining the self- and mutual impedances, even the total number of surrounding metal installations in many practical cases cannot be determined and known. Therefore a completely new and different approach was needed than the one that had been known for a long time.

Also, on the basis of the equivalent circuit in Fig. 3.3, it is not difficult to see that each current induced in the neutral conductors or surrounding metal installations appears as a consequence of the inductive influence of the test current I_t and all other currents induced in remained installations and neutral conductors. Thus each of the presented currents cumulatively involves the inductive influences of all other, known and unknown, currents presented in Fig. 3.3. Also, the value of each of the induced currents depends on the mutual spatial positions (U_{n0}, U_{n1}, U_{n2}, ..., U_{nN}) and the constructive characteristics of all elements (Z_1, Z_2, Z_3, ..., Z_N) of the spontaneously formed electrical circuit, including even those, influence of which is so small that can be neglected. However, these facts are not sufficient for determining the actual current passing through the earth, I_e, and solving the considered problem. They only indicate the direction in which further investigations could be oriented with the aim to enable the compensation of relevant but unknown data. In other words, each of the currents presented in Fig. 3.3 contains the cumulative influence of all relevant factors and parameters, including those that are unavailable. Therefore an important step on the research path was the decision to measure these currents wherever possible, and it is possible only in one of the phase conductors and in the sheaths of the cable line.

Finally, the desired solution was achieved by using measured currents I_t and I_1 and by substituting all surrounding metal installations with only one equivalent conductor, imagined as a cylinder placed around and along the entire length of the considered cable line [1]. Somewhat later, for the sake of simplifying the necessary calculation procedure, it has been shown that this fictitious conductor can be such that can substitute from the standpoint of its inductive influence all sheaths of the considered cable line, including the sheath belonging to the single-core cable used for the simulated ground-fault [3,4].

The presented complete equivalent circuit is not without idealizations and approximations in relation to the actual physical model, although incorporates all, known and unknown, surrounding metal installations. The fact that some of these installations are grounded at certain places, usually at all nearby buildings arranged along the same street(s) as the considered cable line is disregarded. Also, the fact that the metal sheaths in the case of cable lines with applied cross-bonding are grounded at each of the cross-bonding boxes has been disregarded. However, these facts are not at all relevant, because we observe only the currents that are a consequence of inductive couplings. Since they circulate only in the axial direction through these installations, they do not produce any potential on them toward the earth. Thus it is not relevant whether these installations toward the ground are insulated or in direct and continuous contact with the earth along their whole length. Because of the same reason, it is not relevant whether the equivalent conductor substituting all neutral conductors and surrounding metal installations is insulated toward the earth or not. As an example of a similar consideration, we have the well-known case of the determination of the reduction factor of overhead feeding lines. In that case, we also do not take into account the fact that their ground wire(s) are grounded at each line tower, for example [13].

Also, by the circuit in Fig. 3.3 was disregarded the fact that the relevant constructive parameters of some installations are not uniformly distributed along these installations, as well as that all sections of some installations are not completely in parallel in relation to the considered cable line and other of the surrounding metal installations. However, as can be seen in Section 3.5, these data also are not necessary for developing here-presented methodology. These data as well as many others are already cumulatively involved through the measured values of currents I_t and I_1.

Finally, the balanced currents passing through the phase conductors in the conditions of an actual ground fault are not involved in this equivalent circuit. However, the influence of the ground-fault current components passing only through phase conductors on the surrounding metal installations is, as is well known, practically negligible. Something similar can be said for the capacitive currents in the considered cable line during the necessary test measurements. They are disregarded because voltage U_a is very small (usually between 100 and 300 V) in comparison with the nominal voltages of HV and EHV cable lines.

3.2.2 Relevant parameters of the equivalent conductor

As has been shown, each power line passing through urban and/or suburban areas during an actual or simulated ground fault can be represented by a complex electrical circuit as shown in Fig. 3.3. In reality, metal installations representing the elements of this circuit are mainly situated under the surface of the ground or at places that usually are not accessible for performing any measurements. Thus many of the presented currents cannot be determined by direct measurements. Fortunately, this is not the case with the currents passing through the conductors of power lines. This fact enables collecting all relevant data necessary for taking into account the overall inductive influence of surrounding metal installations, including metal sheaths of the two for the measurement purpose superfluous single-core cables, on the distribution of simulated ground-fault current I_t. As will be seen, for this purpose, it is necessary and sufficient to know only two of the currents presented in the complete equivalent circuit in Fig. 3.3. These currents are the simulated ground-fault current through one freely adopted single-core cable I_t and the current in its metal sheath I_1. The measurement circuit necessary for determining these currents is presented in Fig. 3.4.

The used notation has the following meaning: U_a is the auxiliary voltage source; I_t is the simulated ground-fault current through one of the phase conductors of the considered line; $I_1$is the current through the sheath of the cable carrying current I_t; I_i is the current induced in the surrounding metal installations; I_e is the current injected into the surrounding earth through the grounding system of substation B; A is the ampere-meter; and G is the remote ground.

To simplify the complete equivalent circuit presented in Fig. 3.3, let us introduce an equivalent conductor substituting all surrounding metal installations, including metal sheaths of two, for the simulated ground-fault unused single-core cables. On the basis of that the corresponding equivalent circuit of the considered cable line can be presented as shown in Fig. 3.5.

The used notation has the following meaning: U_{01}, (U_{0eq}) is the voltage that current I_1 (I_i) induces in the used phase conductor; U_{10} (U_{1eq}) is the voltage that

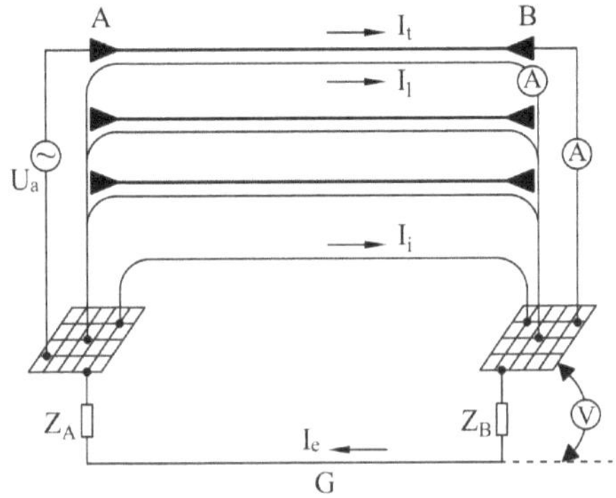

Figure 3.4 The principal measurement circuit.

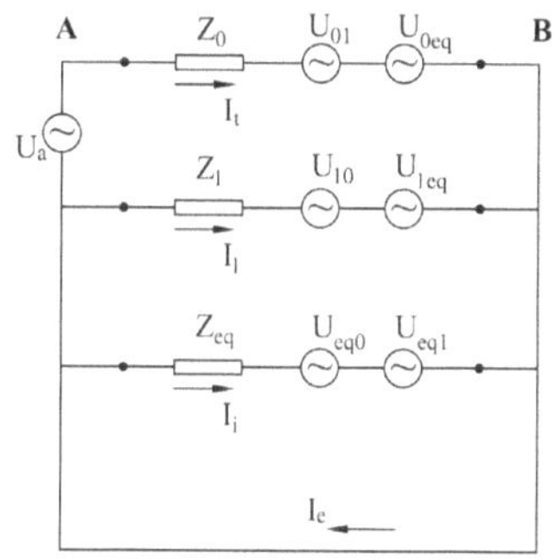

Figure 3.5 Simplified equivalent circuit.

current I_t (I_i) induces in the cable metal sheath; U_{eq0} (U_{eq1}) are the voltages that current I_t (I_1) induced in the introduced equivalent conductor; I_e is the current passing through the earth; and Z_0 is the self-impedance of the used phase conductor.

Concerning voltages, U_{01} and U_{0eq} should be mentioned that their influences are encompassed through the measured value of current I_t so that they are also irrelevant for further analysis.

As can be seen, this circuit is obtained under the assumption that the grounding impedances at the ends of the line are negligibly small, $Z_A \approx Z_B \approx 0$. This assumption is in accordance with the definition of the reduction factor that expresses only the inductive influence between the phase conductors and neutral conductors on the reduction of current dissipated through the grounding system of the supplied substation into the surrounding earth.

By comparing the equivalent circuits presented in Figs. 3.3 and 3.5, it is clear that an arbitrarily large number of unknown currents ($N-2$) has been by introducing an equivalent conductor reduced to only one unknown current, I_i. Also, an arbitrarily large number of known and unknown metal installations surrounding the used single-core cable ($N-3$) are in this way reduced to only one unknown conductor. Thus the arbitrarily large number of unknown, but relevant data is in this way reduced to the relevant data of only one equivalent conductor.

The relevant parameters of the introduced equivalent conductor can be determined by using the condition that currents I_t, I_e, and I_1 have to remain unchanged by the substitution of the equivalent circuit of Fig. 3.3 by the one presented in Fig. 3.5. This condition can also be expressed in the following manner: current I_i in Fig. 3.5 has to be equal to the sum of all currents in Fig. 3.3, except currents: I_t, I_1, and I_e. Under such conditions the required current, I_e, remains unchanged. By using the equivalent circuit in Fig. 3.5, the previously defined condition can be expressed by the following system of equations:

$$
\begin{aligned}
&1.\ Z'_{01}I_t + Z'_1 I_1 + Z'_{1eq}I_i = 0 \\
&2.\ Z'_{0eq}I_t + Z'_{1eq}I_1 + Z'_{eq}I_i = 0
\end{aligned}
\tag{3.5}
$$

where Z'_{eq} is the self-impedance of the introduced equivalent conductor, per unit length; Z'_{0eq} is the mutual impedance between the introduced equivalent conductor and the phase conductor of the single-core cable with test current, per unit length; and Z'_{1eq} is the mutual impedance between the introduced equivalent conductor and the sheath of the single-core cable carrying test current, per unit length.

In the system of Eq. (3.5), impedances Z'_1 and Z'_{01} are determined, according to Ref. [13], by the following expressions

$$Z'_1 = R'_{sh} + \omega_t \frac{\mu_0}{8} + j\omega_t \frac{\mu_0}{2\pi} \ln\frac{\delta_t}{r_{sh}}, \quad \Omega/\text{km}, \tag{3.6}$$

$$Z'_{01} = \omega_t \frac{\mu_0}{8} + j\omega_t \frac{\mu_0}{2\pi} \ln\frac{\delta_t}{r_{sh}}, \quad \Omega/\text{km}. \tag{3.7}$$

where R'_{sh} is resistance of the cable sheath per unit length (Ω/km); and r_{sh} is the mean radius of the cable sheath (m).

With the aim of minimizing the number of unknown but relevant parameters of the introduced equivalent conductor, the physical appearance and spatial position of this conductor are imagined as a cylinder around the single-core cable with current I_t and along the entire cable line length. Under such assumptions, this conductor can also be named equivalent sheath, while the considered cable line is modified into the simple line model, cross section of which is shown in Fig. 3.6.

For the adopted appearance and spatial position of the introduced equivalent conductor, the analytical expressions for impedances Z_{eq}, Z_{0eq}, and Z_{1eq} are a priori known and determined by

$$Z'_{eq} = R'_{eq} + \omega_t \frac{\mu_0}{8} + j\omega_t \frac{\mu_0}{2\pi} \ln\frac{\delta_t}{r_{eq}}, \quad \Omega/\text{km} \tag{3.8}$$

$$Z'_{0eq} = Z'_{1eq} = \omega_t \frac{\mu_0}{8} + j\omega_t \frac{\mu_0}{2\pi} \ln\frac{\delta_t}{r_{eq}}, \quad \Omega/\text{km} \tag{3.9}$$

where R'_{eq} is the resistance of the equivalent sheath per unit length (Ω/km); and r_{eq} is the mean radius of the imagined cylinder representing the equivalent sheath (m).

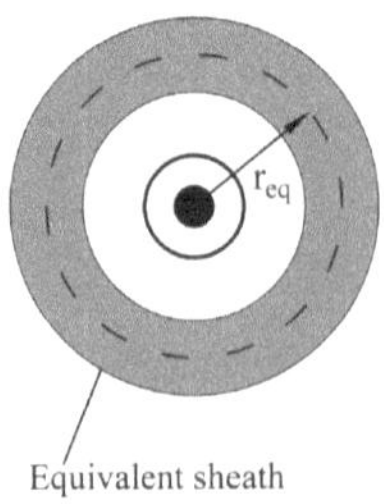

Figure 3.6 The equivalent sheath around a single-core cable.

Since currents I_t and I_1 are obtained by measurements, they can be treated as known quantities, and according to this, instead of Eq. (3.5), we can write

$$\frac{Z'_{0eq}Z'_{1eq} - Z'_{eq}Z'_{01}}{Z'_{eq}Z'_1 - Z'^2_{1eq}} = \frac{I_1}{I_t} \tag{3.10}$$

In this equation the only unknown quantities, according to Eqs. (3.6)–(3.9), are R'_{eq} and r_{eq}. Since Eq. (3.10) gives the relationship between complex quantities, it can be presented as the following system of equations:

$$\begin{aligned} \mathrm{Re}\left\{\left(Z'_{0eq}Z'_{1eq} - Z'_{eq}Z'_{01}\right)I_t\right\} &= \mathrm{Re}\left\{\left(Z'_{eq}Z'_1 - Z'^2_{1eq}\right)I_1\right\} \\ \mathrm{Im}\left\{\left(Z'_{0eq}Z'_{1eq} - Z'_{eq}Z'_{01}\right)I_t\right\} &= \mathrm{Im}\left\{\left(Z'_{eq}Z'_1 - Z'^2_{1eq}\right)I_1\right\} \end{aligned} \tag{3.11}$$

After solving this system of equations and determining the relevant parameters of the introduced equivalent conductor (R'_{eq} and r_{eq}), relations Eqs. (3.8) and (3.9) can be used for obtaining Z_{eq}, Z_{0eq}, and Z_{1eq}, as well as currents I_i and I_e, by using the equivalent circuit of Fig. 3.5. Then, on the basis of this equivalent circuit and the reduction factor definition, the actual feeding line reduction factor is determined by

$$r = \frac{I_e}{I_t} = 1 - \frac{\left(Z'_{eq} - Z'_{1eq}\right)Z'_{01} + \left(Z'_1 - Z'_{1eq}\right)Z'_{0eq}}{Z'_1Z'_{eq} - Z'^2_{1eq}}, \tag{3.12}$$

or, according to Eqs. (3.6) and (3.7), in a somewhat more compact form

$$r = 1 - \frac{R'_{eq}Z'_{01} + \left(Z'_1 - Z'_{1eq}\right)Z'_{0eq}}{Z'_1Z_{eq} - Z'^2_{1eq}}. \tag{3.13}$$

In the case of the line consisting of three single-core cables, the reactive parts of impedances Z_1 and Z_{1eq} are approximately equal ($X_1 \approx X_{1eq}$) and, instead of Eq. (3.13), the following approximation can be used:

$$r \approx 1 - \frac{R'_{eq}Z'_{01} + R_1Z'_{0eq}}{Z'_1Z'_{eq} - Z'^2_{1eq}}. \tag{3.14}$$

Finally, if one assumes that the cable sheaths do not exist ($Z_1 \rightarrow \infty$, $R_1 \rightarrow \infty$, $Z_{01} = 0$ and $Z_{1eq} = 0$), the line reduction factor, existing only as a consequence of the influence of surrounding metal installations, is

$$r_{mi} \approx \frac{Z'_{eq} - Z'_{0eq}}{Z'_{eq}}, \tag{3.15}$$

or, in the more direct form:

$$r_{mi} \approx \frac{R'_{eq}}{R'_{eq} + \omega_t\left(\mu_0/8\right) + j\left(\omega_t\mu_0/2\pi\right)\ln\left(\delta_t/r_{eq}\right)}. \tag{3.16}$$

Here, it is interesting to mention that thanking the imagined physical appearance and spatial position of the introduced equivalent conductor, the number of its parameters unknown but relevant for solving the considered problem is reduced to only two of them. For any other freely imagined appearance and spatial position of this conductor, the considered problem is not solvable because the number of its parameters that should be determined would be greater than the number of equations that can be written.

On the basis of the previously presented analytical procedure and equivalent circuit of Fig. 3.5, it is not difficult to show that the relevant parameters of this equivalent conductor are determined by

$$R'_{eq} = \mathrm{Re}\left\{Z'_1 Z'_{eq} - Z'^2_{1eq}\right\}. \tag{3.17}$$

$$r_{eq} = \frac{\delta}{e^{2\pi X'_{eq}/\omega\mu_0}} \tag{3.18}$$

where R'_{eq} and X'_{eq} are the real and imaginary parts of the longitudinal self-impedance of the introduced equivalent conductor.

Also, based on the determined parameters R'_{eq} and r_{eq}, as well as the equivalent circuit in Fig. 3.5, one can determine the parameters of the equivalent sheath substituting all cable line sheaths and all surrounding metal installations. Then the considered cable line during a ground fault in the supplied substation can be presented by a very simple cable line model shown in Fig. 3.7.

Concerning used approximations $Z_A \approx 0$ and $Z_B \approx 0$ should be mentioned that HV distribution substations located in urban or suburban areas are surrounded by many underground metal installations that act as perfect grounding electrodes [12,14]. Thus the quantitative relationships between the impedances in the equivalent circuit in Fig. 3.3 are such that grounding impedances Z_A and Z_B can be disregarded, which is following the assumption used for the definition of the reduction factor of feeding lines (see Chapter 2: Calculated reduction factor of feeding cable lines). On the basis of these approximations, it can be said that each current appearing in any of the line neutral conductors or any of the surrounding metal installations is a consequence solely of the mutual inductive coupling with each of all

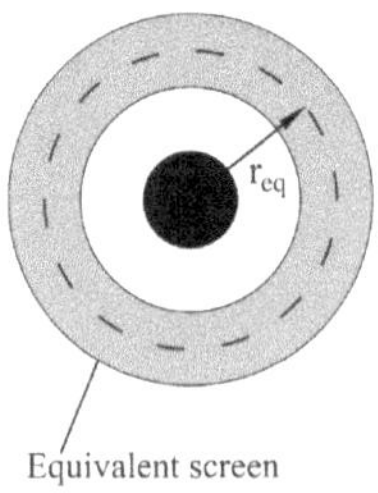

Figure 3.7 The equivalent sheath surrounding the phase conductor with current I_t.

other surrounding conductors and installations (Fig. 3.3). However, the presented method is developed based on the measured I_t and I_1. According to that it takes into account the fact that the real values of Z_A and Z_B are somewhat larger than zero. Because of that currents I_t and I_1 contain components appearing only as a consequence of potentials $Z_A I_e$ and $Z_B I_e$ (Figs. 3.3 and 3.4). However, these components of currents I_t and I_1 are negligibly small in practical conditions because of the following facts:

- current I_e is only a few percent of I_t because of strong inductive coupling between the phase conductor with current I_t on the one side and the cable line sheaths, including all surrounding metal installations, on the other side, and
- the length of HV or EHV cable lines is such that almost each HV distribution substation and its feeding cable line satisfy the following relation:

$$Z_A \approx Z_B < Z'_{ph}\mathrm{L}, \tag{3.19}$$

where L is the total length of the considered HV cable line.

Since the spatial positions of the individual single-core cables in relation to the surrounding metal installations are not identical or symmetrical, the presented method should be applied for each of the line phase conductors separately. However, for the estimation of safety conditions within and in the vicinity of a supplied substation, we should adopt as a critical the equivalent sheath parameters that correspond to the smallest inductive influence of the surrounding metal installations. Certainly, in this way, the finally obtained value of the reduction factor is the most critical from the standpoint of safety conditions within and in the vicinity of a supplied substation.

Also, the value of equivalent soil resistivity ρ in urban areas can only be approximately estimated on the basis of the main geological characteristics of the soil surrounding the considered power line. However, according to Eqs. (3.1) and (3.2), the favorable circumstance is the fact that impedances Z_n and Z_{nm} are only slightly dependent on this parameter. Also, it is necessary to mention here that within the whole approximately estimated range of actually possible values of ρ, the lowest one should be adopted for necessary calculations because it gives a smaller inductive influence, that is, reduction factor value that is slightly larger than its real value.

Finally, the presented methodology version is not applicable in the cases of HV or EHV cable lines with applied cross-bonding, which should be, according to the corresponding technical standards, for example [15], the most common case in contemporary power distribution practice.

3.2.3 *Critical ground-fault position*

The described method enables us to obtain the value of the actual reduction factor of a power line surrounded by an arbitrary number of metal installations when a ground fault occurs in the supplied substation itself. However, whether this fault place is critical from the standpoint of safety conditions in the supplied substation, that is, whether this fault place causes the largest value of the ground-fault current fraction passing through the grounding system of this substation?

Total ground-fault current for any fault place along an HV or EHV lane is, according to Ref. [13], determined by the following equation:

$$I_{FA}(l) = \frac{3U_{ph}}{Z_A^+ + Z_A^- + Z_A^0 + l\left(2z_l^+ + z_l^0\right)} \tag{3.20}$$

where U_{ph} is the phase voltage at supply substation A (kV); I_{FA} is the ground-fault current component coming from substation A (kA); Z_A^+, Z_A^-, and Z_A^0 are the positive, negative, and zero sequence impedance of the power system at substation A (Ω); z_l^+, and z_l^- are the positive and zero sequences impedance of the cable line, per unit length (Ω/km); and l is the cable line length from supply substation A to fault place F (km).

Also, based on Fig. 3.8 (see Chapter 2: Calculated reduction factor of feeding cable lines), it is not difficult to see the following: relative ground-fault current distribution at different fault places along an HV or EHV cable line is independent of the total number of the neutral conductors (cable line sheaths). Since each surrounding metal installation grounded at its ends acts as a specific neutral line conductor, it can be said that relative ground-fault current distribution at different places along an HV or EHV cable line is not dependent on the total number of cable line sheaths and surrounding metal installations, N in Fig. 3.3. Also, the inductive influence of all cable line sheaths and surrounding metal installations on the way of ground-fault current distribution is expressed only through the value of the reduction factor in the case when a ground fault occurs in the supplied substation. On the basis of these observations and the analytical expression (2.23), the reduction factor for a fault anywhere along one cable line is a linear function of the line length to the fault place, or

$$r_a(l) = \frac{I_e(l)}{I_{FA}(l)} = \frac{l}{L} r_a, \tag{3.21}$$

where $r_a(l)$ is the actual reduction factor for a fault at an arbitrary distance l along the line, seen from substation A; r_a is the actual cable line reduction factor for a fault in supplied substation B; and L is the total length of a cable line (km).

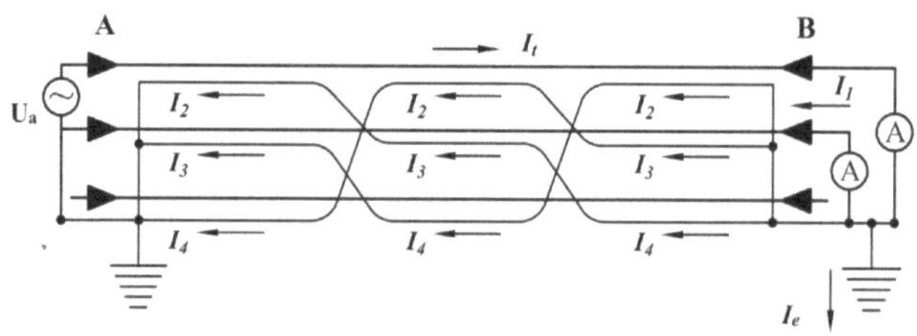

Figure 3.8 The principal measurement circuit.

Since the actual cable line reduction factor is smaller than the calculated one ($r_a < r$), reduction factor $r_a(l)$ is, according to Eqs. (3.23) and (3.21), in the same relationship smaller than reduction factor $r(l)$ in any point along any HV or EHV cable line.

On the basis of Eqs. (3.20) and (3.21), it can be seen that the ground-fault current fraction flowing through the earth for a fault anywhere along an HV or EHV line is determined by

$$I_e(l) = r_a(l)I_{FA}(l) = \frac{l}{L}\frac{3U_{ph}}{Z_A^+ + Z_A^- + Z_A^0 + l\left(2z_l^+ + z_l^0\right)}r_a \tag{3.22}$$

Since impedances Z_A^+, Z_A^-, and Z_A^0 are always somewhat larger than zero, based on Eq. (3.22), it is clear that current $I_e(l)$ has its largest/critical value when a fault occurs at the end of the feeding cable line. Thus cable metal sheaths and surrounding metal installations grounded at their ends have no influence on critical, from the standpoint of safety conditions within and in the vicinity of the supplied substation, fault place. This place is in all cases, including the longitudinally inhomogeneous lines that will be considered in Chapter 4, Reduction factor of lines consisting of one overhead and one cable section and Chapter 5, Measures for improving reduction factor of a feeding line, in the supplied substation itself.

On the basis of similar analysis, it is not difficult to see that the critical fault position from the standpoint of thermal stress in the cable sheaths is somewhere nearby supply substation A. Then the total ground-fault current is critical from the standpoint of thermal stress and is, according to Eq. (3.20), determined by the following approximation:

$$I_{FA}(l \approx 0) \approx \frac{3U_{ph}}{Z_A^+ + Z_A^- + Z_A^0} \tag{3.23}$$

3.3 Cable lines with applied cross-bonding

3.3.1 Measurement procedure

The length of HV and EHV cable lines in the distribution network is so large that the application of the cross-bonding technique is required for each of them [15]. With the aim to eliminate harmful circulating currents through the sheaths of HV or EHV cable lines, the route of these lines is divided into major sections, each of which is divided into three subsections. In each of these subsections, the three sheaths connected in series are associated with conductors of different phases, and their currents, and consequently the voltages of the sheaths have the same magnitude but with a phase shift of 120 degrees. The overall effect is that the resultant voltage and the current in each of the sheaths are approximately equal to zero. However, at the same time, cable sheaths lose their screening effects during normal operation conditions, and mutual inductive influence between phase conductors and the surrounding metal installations as well as between the surrounding metal installations themselves becomes intensified, which will be considered in Chapter 7, Normal operation conditions and chapter 8, Overhead lines and cable lines with no applied cross-bonding.

Due to the applied cross-bonding, the analytical expression for the mutual impedance between the phase conductors and their sheaths becomes more complex and as such is not suitable for the application of the methodology version presented in Section 3.2.2. Besides, the cable line sheaths are grounded at each cross-bonding box along the line and because of that current I_1 would not be only the consequence of inductive coupling between the phase conductor with current I_t and the cable sheath with current I_1. As such this current would not be applicable for the determination of the equivalent sheath parameters.

The favorable circumstance is that the mutual spatial positions of all phase conductors remain unchanged along the entire length of cable lines with cross-bonding. Accordingly, the analytical expression for their mutual impedances has a relatively simple mathematical form. Besides, since these conductors are not grounded at any place along the line, the currents induced in these conductors have a constant value along the whole line length. Thanking this, measurement of an induced current, I_1, is possible at the accessible places, that is, in the supply or supplied substation. As an illustration mutual spatial positions of phase conductors, that is, single-core cables laid in flat formation along the whole line length are shown in Fig. 3.2.

By using the mentioned favorable circumstance, we will here, instead of the cable sheath, use one more of the phase conductors (in the further consideration named the "additional phase conductor") for the necessary test measurements. On the basis of that the necessary measurement circuit here is somewhat modified as is shown in Fig. 3.8.

The used notation has the following meaning: U_a is the auxiliary voltage source; I_t is the test (simulated) ground-fault current; I_1 is the current induced in the additional phase conductor; I_e is the current injected into the earth through the grounding system of substation B; I_2, I_3, and I_4 are the currents through the individual cable line sheaths; and A is the ampere-meter.

According to the previous considerations, the complete equivalent circuit presented in Fig. 3.3, in this case, should be enlarged by one more phase conductor acting as one more neutral conductor and used for obtaining induced current I_1.

When we introduce an equivalent neutral conductor as well as in the previous case (Section 3.2) that substitutes all surrounding metal installations and cable line sheaths, the considered cable line during the necessary measurements can be presented by a relatively simple line model, cross section of which is shown in Fig. 3.9.

The additional phase conductor is marked differently (lighter shaded) to show that for the measurement purpose, it serves as one neutral conductor. In the here-considered case the equivalent sheath, according to Fig. 3.8, encompasses beside the inductive influences of all surrounding metal installations also the inductive influences of all sheaths of the considered line.

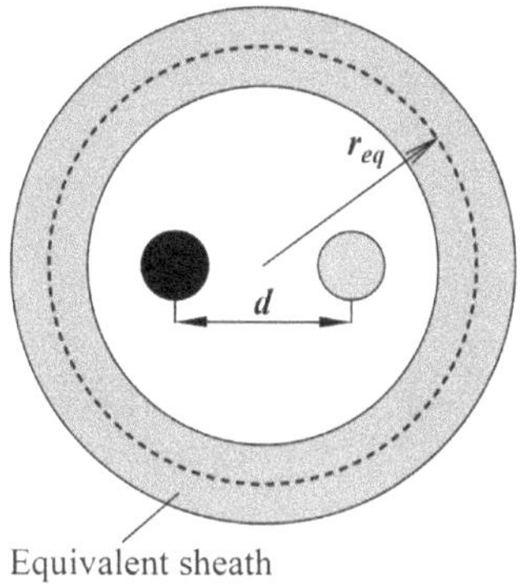

Figure 3.9 Simplified line model during the measurements.

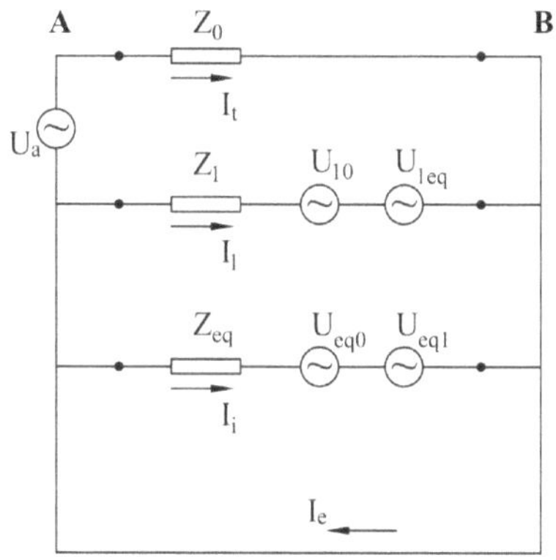

Figure 3.10 Equivalent circuit of the simplified line model shown in Fig. 3.9.

According to the circuit shown in Fig. 3.8 and the simplified line model in Fig. 3.9, during the test measurements, the considered line consists of one phase conductor and two neutral conductors. One of them is the additional phase conductor, while the other is the introduced equivalent sheath. Thus the corresponding equivalent circuit of the line prepared for test measurements is relatively simple as is shown in Fig. 3.10.

As can be seen in the presented equivalent circuit impedances Z_A and Z_B are disregarded ($Z_A \approx Z_B \approx 0$), which is following the definition of the reduction factor of feeding lines.

3.3.2 Relevant parameters of the equivalent sheath

On the basis of the equivalent circuits in Figs. 3.3 and 3.10, it is obvious that an arbitrary large number of unknown currents (N) through the surrounding metal installations and sheaths of the considered cable line is substituted by only one unknown current, I_i. Also, numerous known and unknown surrounding metal installations, including sheaths of the considered line, are substituted by only one equivalent conductor. The relevant parameters of this conductor can be determined by using the condition that the currents I_t, I_1, and I_e have to remain unchanged after this substitution (Figs. 3.3 and 3.10). This condition can be expressed by the following system of equations:

$$\begin{aligned} &1.\ \ Z'_{01}I_t + Z'_{ph}I_1 + Z'_{1eq}I_i = 0 \\ &2.\ \ Z'_{0eq}I_t + Z'_{1eq}I_1 + Z'_{eq}I_i = 0 \end{aligned} \tag{3.24}$$

where Z'_{eq} is the self-impedance of the introduced equivalent sheath, Ω/km; Z'_{0eq} is the mutual impedance between the introduced equivalent sheath and the phase conductor with current I_t, Ω/km; Z'_{1eq} is the mutual impedance between the introduced equivalent sheath and the additional phase conductor, Ω/km.

Impedances Z'_{ph} and Z'_{01} are, according to Ref. [13], determined by

$$Z'_{ph} = R'_{ph} + \omega_t \frac{\mu_0}{8} + j\omega_t \frac{\mu_0}{2\pi}\left(\frac{\mu_r}{4} + \ln\frac{\delta_t}{r_{ph}}\right), \quad \Omega/\text{km}, \tag{3.25}$$

$$Z'_{01} = \omega_t \frac{\mu_0}{8} + j\omega_t \frac{\mu_0}{2\pi}\ln\frac{\delta_t}{d}, \quad \Omega/\text{km}, \tag{3.26}$$

where R'_{ph} is the longitudinal resistance of the phase conductor (Ω/km); r_{ph} is the radius of the phase conductor (m); ω_t is the angular test circuit frequency: $2\pi f_t$; and d is the distance between two adjacent single-core cables (Fig. 3.2).

With the aim of minimizing the number of relevant but unknown parameters of the introduced equivalent conductor, the physical appearance and space position of this conductor are imagined as a cylinder around the phase conductors with currents I_t and I_1 along the entire line length, Fig. 3.9. In this way the analytical expressions for impedances Z_{eq}, Z_{0eq}, and Z_{1eq} become a priori known and, according to Ref. [13], determined by Eqs. (3.8) and (3.9).

Also, since currents I_t and I_1 are obtained by measurements, they can be treated as known quantities, and Eq. (3.24) can be modified into

$$\frac{Z'_{0eq}Z'_{1eq} - Z'_{eq}Z'_{01}}{Z'_{eq}Z'_{ph} - Z'^2_{1eq}} = \frac{I_1}{I_t}. \tag{3.27}$$

Finally, since expression (3.27) gives the relationship between complex quantities, it can be presented as a system of the following two equations:

$$\begin{aligned} &\text{Re}\left\{\left(Z'_{0eq}Z'_{1eq} - Z'_{eq}Z'_{01}\right)I_t\right\} = \text{Re}\left\{\left(Z'_{eq}Z'_{ph} - Z'^2_{1eq}\right)I_1\right\} \\ &\text{Im}\left\{\left(Z'_{0eq}Z'_{1eq} - Z'_{eq}Z'_{01}\right)I_t\right\} = \text{Im}\left\{\left(Z'_{eq}Z'_{ph} - Z'^2_{1eq}\right)I_1\right\}. \end{aligned} \tag{3.28}$$

where Re and Im are the real and imaginary parts of Eq. (3.28).

In Eq. (3.28) the only unknown quantities, according to Eqs. (3.8), (3.9), (3.25), and (3.26), are R'_{eq} and r_{eq}. Thus by solving these equations, we can determine the required parameters, R'_{eq} and r_{eq}.

When the parameters of the introduced equivalent sheath are known, we can present each cable line with applied cross-bonding during a ground fault in the supplied substation by the simplest cable line consisting of only one single-core cable as shown in Fig. 3.7. Thus for obtaining the actual reduction factor of any cable line with applied cross-bonding, we can use the analytical expression for the reduction factor of the cable line constructed of only one single-core cable. Thus according to Eq. (3.10), we obtain

$$r = \frac{R'_{eq}}{R'_{eq} + \omega(\mu_0/8) + j(\omega\mu_0/2\pi)\ln(\delta/r_{eq})}. \tag{3.29}$$

Because of the previously introduced approximations: $Z_A \approx Z_B \approx 0$, it is necessary to remind that HV distribution substations are located in urban or suburban areas having many underground metal installations that act as perfect grounding electrodes and/or as conductive connections with other grounding electrodes, like foundation earth electrodes of surrounding buildings [14]. Thus the quantitative relationships between the impedances shown in the complete equivalent circuit (Fig. 3.3) are such that the influence of grounding impedances Z_A and Z_B can be disregarded ($Z_A \approx 0$ and $Z_B \approx 0$). On the basis of these approximations, each current appearing in the additional phase conductor or any of the surrounding electrical circuits is a consequence solely of the mutual inductive couplings each of the electrical circuits with all other surrounding electrical circuits (Figs. 3.3 and 3.8). However, the development of the presented methodology is based on the measured currents I_t and I_1. According to this, it takes into account the fact that the real values of Z_A and Z_B are somewhat larger than zero. Because of that currents I_t and I_1 contain components appearing only as a consequence of potentials Z_AI_e and Z_BI_e (Fig. 3.8). However, these components of currents I_t and I_1 are negligibly small in practical conditions because of the following facts:

- current I_e is only a few percent of I_t because of the high conductivity of the additional phase conductor as well as because of strong inductive coupling between the phase conductor with current I_t on the one side and the additional phase conductor, including all surrounding metal installations and cable sheaths on the other side, and
- the actual length of HV cable lines in an HV distribution network is such that the following inequality is almost always, according to for example, Refs. [12,14], satisfied:

$$Z_A \approx Z_B < Z'_{ph}L, \tag{3.30}$$

where L is the total length of an HV or EHV cable line.

Since the spatial positions of the individual single-core cables in relation to the surrounding metal installations are not identical, the presented procedure should be applied for each of the line phase conductors. After that, for the safety conditions assessment, one adopts those parameters (R'_{eq} and r_{eq}) that correspond to the phase conductor with the smallest inductive influence of the surrounding metal installations. Certainly, in this way, the finally determined actual reduction factor is slightly on the side of enlarged safety within and in the vicinity of the supplied HV substation.

As is shown the here-described methodological version is based on using one more phase conductor of the considered line for the necessary test measurements. Thanks to this, it is applicable for any other HV or EHV line, for example, cable lines with no cross-bonding, overhead lines, and lines constructed as a combination of one cable and one overhead section (the methodological version that uses a neutral conductor cannot be applied in this case). Also, this version is easy for practical application and more accurate because of the greater conductivity of phase conductors in comparison with neutral conductors of power lines. Having in mind these facts concerning the presented methodological version, it can be said that the posed goal of the presented research procedure has been completely achieved. The presented methodological version enables the determination of the actual reduction factor of HV and EHV lines passing through urban and/or suburban areas by taking into account the favorable inductive influence of all surrounding metal installations. Given this, the previous methodological versions have no practical significance and should be viewed only as stages in the research process that led to the ultimate goal.

The surrounding metal installations have a very pronounced favorable influence on the distribution of ground-fault current in HV substations located in urban or suburban areas. A certain insight into the quantitative side of the considered influences can be obtained on the basis of the results of experimental measurements and calculations presented in Section 3.6.

3.4 Overhead lines

In contemporary HV distribution networks, overhead lines are also applied, especially in areas of a lower degree of urbanization. Because of that, in the case of an overhead line passing through urban and/or suburban areas, its actual reduction factor of this type of HV and EHV distribution lines also cannot be determined on the basis of the existing analytical expressions [13].

In this case the number of neutral conductors is usually only one and, according to this, the complete equivalent circuit presented in Fig. 3.3 for a cable line, in this case, is somewhat simpler. However, the data concerning the surrounding metal installations are also uncertain or completely unknown and because of that, for the sake of an illustration of currents induced in these installations, we can also use the equivalent circuit shown in Fig. 3.3.

Also, in comparison with the case considered in Section 3.2.1, instead of the two cable line phase conductors, in this case, two phase conductors of an overhead line are used for necessary measurements. Since the analogy is obvious, the same measurement and calculation procedure can be used in this case like the one described in the previous section. At this, one assumes that the introduced equivalent neutral conductor has also a cylindrical shape and surrounds both phase conductors used for obtaining currents I_t and I_1, as shown in Fig. 3.9. On the basis of these assumptions, the considered overhead line and all surrounding metal installations during a simulated ground fault can also be presented by a simple physical model, cross section of which is shown in Fig. 3.7, while its equivalent circuit is shown in Fig. 3.10. When one obtains relevant parameters of the introduced equivalent neutral conductor, the actual reduction factor of this type of distribution line can be determined by using a procedure analogous to the one described in the previous section.

Here, it is also important to point out that the portions of current I_t appearing in the adopted phase conductors as a consequence of the potentials of the grounding electrodes at the line ends (Z_AI_e and Z_BI_e) are, because of the significantly greater mutual distances between phase conductors, somewhat larger than in the previous case. However, because of the great conductivity of phase conductors and usually the great length of overhead distribution lines, the influence of these components of currents I_t and I_1 on the methodology accuracy, in this case, is also negligible.

Because of the significantly greater mutual distance between the phase conductors and their greater asymmetry toward the neutral conductor(s) and surrounding metal installations, it is necessary to repeat the whole presented procedure for each phase conductor separately. The critical one, from the standpoint of the safety conditions in the supplied substation, is certainly the phase conductor resulting in the largest value of current I_e.

3.5 Theoretical foundations for introducing equivalent sheath

3.5.1 Electromagnetic interpretation of the methodology

The introduced equivalent conductor in the presented analytical procedure can be explained in a way that is somewhat nearer to the actual physical phenomenon, known as electromagnetic induction. Namely, each of the cable line sheaths and each of the surrounding metal installations acts as a short-circuited "secondary turn" of one spontaneously formed and specific transformer. By using this analogy an HV or EHV cable line during the test measurements (Figs. 3.3 and 3.9) can be, with certain idealizations and approximations, presented in the way shown in Fig. 3.11.

The used notation has the following meaning: F_r is the resultant magnetic flux through the "primary winding," and G is the ground.

On the basis of the presented physical model, the previously used condition (currents I_t, I_1, and I_e should remain unchanged after introducing an equivalent conductor) can also be defined in the following manner: the resulting alternating magnetic

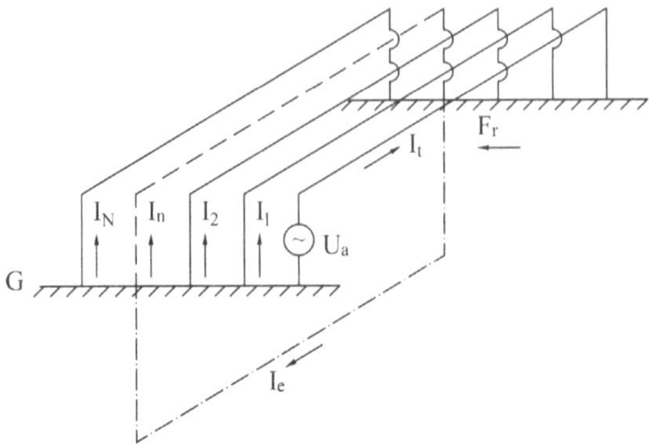

Figure 3.11 Schematic presentation of the actual physical model.

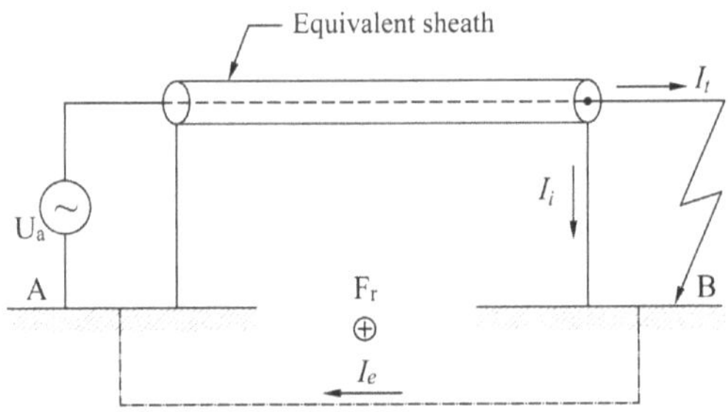

Figure 3.12 Actual model simplified by the equivalent sheath.

flux through the "primary winding" (the surface closed by the phase conductor with current I_t and the earth as a common return path), F_r, should remain unchanged after introducing an equivalent conductor.

Also, based on Fig. 3.11, one can notice that the equivalent conductor can be determined for any, including the irregular geometric shape of some "secondary turn" (one of surrounding metal installations, including the earth as a common return path) as well as for an arbitrarily large number of such turns, N. The same can be concluded concerning the geometric shape of the "primary turn," as well as concerning the turn formed by the introduced equivalent conductor (equivalent secondary turn). For a certain value of current I_t, the magnitude of the resulting flux depends only on the sizes of the surface covered by these spontaneously formed turns and their positions toward the "primary winding (turn)," whereas their geometrical shapes are not relevant. However, to obtain a smaller number of unknown parameters of the introduced equivalent conductor and solve the problem, this conductor is assumed as a cylinder along and around the considered phase conductor of a tested line as is shown in Fig. 3.12.

On the basis of Fig. 3.12 and the general definition of magnetic flux, it is not difficult to see that F_r depends on the spatial position of the equivalent sheath (equivalent secondary turn) in relation to the "primary turn," r_{eq} (mean radius of the cylinder representing the equivalent conductor), and current through this conductor, I_i. Thus it is not difficult to see that current I_i depends only on r_{eq} and R'_{eq}, for the identical all other relevant factors and parameters. According to this fact, it is not difficult to conclude that F_r also depends only on r_{eq} and R'_{eq}, for the unchanged all other relevant conditions and circumstances. Concerning the other relevant conditions and circumstances, it is interesting to note that through measured currents I_t and I_1, the presented methodology takes into account all their known and unknown relevant data. For instance, through measured I_t and I_1 we take into account, inter alia, all geological characteristics of the surrounding soil, as they really are, including the fact that the soil structure in practical conditions is always heterogeneous.

3.5.2 Cable sheath parameter expressing the influence of surrounding metal installations

The system of Eq. (3.28) represents an undefined system of equations with an infinite number of solutions for r_{eq} and corresponding R'_{eq}. This means that the mentioned condition is satisfied for an arbitrarily large number of different combinations of r_{eq} and corresponding R'_{eq}. All of this can easily be seen only on the basis of the actual physical model of a power line and its analogy with one specific transformer shown in Fig. 3.11. Namely, an arbitrarily large number of the mutually different "equivalent secondary turn" (mutually different combinations of relevant parameters r_{eq} and R'_{eq}), seen individually, can create one the same value of flux F_r and satisfy the condition required for the introduction of one equivalent conductor (secondary turn).

According to the previous consideration, for any a priori adopted the value of r_{eq}, we can determine corresponding R'_{eq} satisfying the required condition. At this, for obtaining the corresponding R'_{eq}, one can use the expression (3.29) in which the value of reduction factor of the line, r, is a priory known quantity determined by the here-presented procedure.

Also, here it is interesting to note one more possibility of the here-presented methodology of compensation of deficiency of an arbitrarily large number of relevant data. The inductive influence of all surrounding metal installations, including cable line sheaths, can be expressed through only one equivalent parameter of the actual, from the standpoint of its physical appearance, sheath of a single-core cable with current I_t. This becomes possible, when we adopt that spatial position of the equivalent cable sheath coincides with the actual cable sheath, $r_{eq} = r_{sh}$ (mean radius of the actual cable sheath). Then, instead of the equivalent parameters r_{eq} and R'_{eq} we have only one equivalent parameter, longitudinal resistance of cable sheath, $R'_{eq}(r_{eq} = r_{sh})$, substituting, from the standpoint of the inductive influence, surrounding metal installations, including sheaths of the considered cable line. Then we have only one cable sheath, dimensions and spatial position of which would remain unchanged. Only its longitudinal resistance is changed in such a way that the value of this parameter expresses the inductive influence of all surrounding metal installations, including cable sheaths of the considered line.

Such an equivalent longitudinal resistance of cable sheath can be determined for each single-core cable of any HV or EHV cable line, including those with applied cross-bonding technique. In this way the physical appearance of the whole cable line remains unchanged. This possibility is utilized for determining the inductive influence of a cable line during normal operating conditions as shown in Chapter 7, Normal operation conditions.

Based on the previous consideration, we are coming to the following, very interesting cognition:

- The inductive influences of all surrounding metal installations, including cable line sheaths in any practically possible situations, can be expressed only through an equivalent longitudinal resistance of the cable sheath of one single-core cable. This equivalent parameter is determined by an imaginary number, effective value of which is smaller than the actual longitudinal resistance.

Of course, similar consideration can be performed to a more complex situation when instead of one we would have more phase conductors that create a magnetic flux through a common contour.

3.5.3 Mathematical interpretation of the methodology

Based on the complete equivalent circuit of an HV or EHV cable line given in Fig. 3.3, we can write the following system of equations:

$$\begin{aligned}
&Z_{ph}I_t + Z_{01}I_1 + Z_{02}I_2 + \cdots + Z_{0n}I_n + \cdots + Z_{0N}I_N = U_a \\
&Z_{10}I_t + Z_{11}I_1 + Z_{12}I_2 + \cdots + Z_{1n}I_n + \cdots + Z_{1N}I_N = 0 \\
&Z_{20}I_t + Z_{21}I_1 + Z_{22}I_2 + \cdots + Z_{2n}I_n + \cdots + Z_{2N}I_N = 0 \\
&\vdots \\
&Z_{n0}I_t + Z_{n1}I_1 + Z_{n2}I_2 + \cdots + Z_{nn}I_n + \cdots + Z_{nN}I_N = 0 \\
&\vdots \\
&Z_{N0}I_t + Z_{N1}I_1 + Z_{N2}I_2 + \cdots + Z_{nN}I_n + \cdots + Z_{NN}I_N = 0
\end{aligned} \tag{3.31}$$

Before the foreseen test measurements, in the given system of equations only the impedances Z_{ph}, Z_{11}, Z_{22}, Z_{33}, Z_{01}, Z_{02}, Z_{03}, Z_{12}, Z_{13}, Z_{23} can be obtained by calculations and can be considered in advance known quantities together with voltage U_a. After the foreseen measurements and obtained values for currents I_t and I_1, as unknown quantities remain current I_i and all other impedances until Z_{NN}. However, after introducing the equivalent neutral conductor substituting all surrounding metal installation, including metal sheaths of the considered line, the number of unknown quantities is significantly reduced so that we obtain the following system of only three equations:

$$\begin{aligned}
&1.\ \ Z_{ph}I_t + Z_{01}I_1 + Z_{0eq}I_i = U_a \\
&2.\ \ Z_{10}I_t + Z_{11}I_1 + Z_{1eq}I_i = 0 \\
&3.\ \ Z_{eq0}I_t + Z_{eq1}I_1 + Z_{eq}I_i = 0.
\end{aligned} \tag{3.32}$$

where

$$I_i = I_2 + I_3 + I_4 + \cdots + I_N. \tag{3.33}$$

In the system of Eq. (3.32) the unknown quantities are I_i, Z_{0eq}, Z_{1eq}, and Z_{eq}, and since the number of unknown quantities is greater than the number of equations this system of equations seen generally is unsolvable. However, by introducing an equivalent conductor of such physical appearance and spatial position that it represents a cylinder that encompasses the phase conductors with currents I_t and I_1 along their entire length, the impedances Z_{0eq} and Z_{1eq} become equal to each other and the number of unknown quantities becomes smaller. Only under such assumptions concerning the introduced equivalent conductor, we obtain the solvable system of equations, Eq. (3.28), and all required quantities, I_i, Z_{0eq}, and Z_{eq} can be determined. When impedances Z_{0eq} and Z_{eq} are determined, it is not difficult by using expressions (3.8) and (3.9) to determine the relevant parameters of the introduced equivalent conductor/sheath, R'_{eq}, and r_{eq}, as well as current I_i. Here it is interesting to note that any other imagined appearance of the introduced equivalent conductor does not give us the desired solution to the considered problem.

As will be shown in Chapter 7, Normal operation conditions, by using a similar procedure, it is possible to determine any (mth) current in the system of Eq. (3.31), I_m. It is only necessary that during the necessary measurements, the installation through which this current would flow is ungrounded at one of its ends, that is, that there is no current in this installation. If the spatial position of the installation in relation to the phase conductor with current I_t is known, when we determined the relevant parameters of the equivalent cable sheath without the participation of this installation, required current I_m can be determined by using the following system of equations:

$$\begin{aligned} &1.\quad Z_{ph}I_t + Z_{0eq}I_i + Z_{0eq}I_m = U_a \\ &2.\quad Z_{10}I_t + Z_{1eq}I_i + Z_{1eq}I_m = 0 \\ &3.\quad Z_{eq0}I_t + Z_{eq1}I_i + Z_{eq}I_m = 0. \end{aligned} \tag{3.34}$$

where

$$I_i = \sum_{n=2}^{n=N} I_n, \quad n \neq m. \tag{3.35}$$

By introducing the equivalent conductor substituting all three phases conductors, the problem of determining and R'_{eq} and r_{eq} can also be solved in cases where the test current, I_t, passes through all three phases conductors as will be shown in Chapter 8, Overhead lines and cable lines with no applied cross-bonding. Also, in cases when load currents flow through phase conductors located at a somewhat greater distance from each other, the problem of determining the currents induced in any of the surrounding metal installations, seen separately, can be solved using the principle of superposition as shown in Chapter 7, Normal operation conditions.

When the line is in normal operating conditions, then the surrounding metal installations affect the load currents in all three phase conductors so that to determine the parameters of the equivalent neutral conductor, a test current is required in all phase conductors if they are placed on a greater mutual distance. Therefore to obtain induced current I_1, it is not possible to use any of them and the measurement procedures described in Sections 3.2 and 3.3. Then, the necessary measurements should be performed by using the measurement circuit shown in Chapter 8, Overhead lines and cable lines with no applied cross-bonding, Figure 8.4. However, in the case of cable lines, the mutual distance of single-core cables is relatively small in relation to their distance from the surrounding metal installations. Due to this, in the case of cable lines, the selection of the phase conductor for obtaining the measurement circuit in Fig. 3.8 has no significant influence on the values of the relevant parameters of the equivalent cable sheath.

3.6 Quantitative analysis

The grounding problem of HV/MV substations located in urban and suburban areas can be under certain conditions and circumstances very pronounced. These conditions and circumstances can be too high local soil resistivity and/or a high level of the short-circuit currents, immediate vicinity of public buildings and objects, and a strictly defined and cramped plot foreseen for laying a sufficiently large grounding electrode. In such conditions and circumstances, it is of prime importance to take into account the favorable influences of the underground metal installations that exist in the urban surroundings independent of our need for solving this problem. Some of these installations are located around HV/MV distribution substations and significantly reduces the impedance of their spontaneously formed grounding system, while the others are located along and around their feeding line and significantly reduce the portion of ground-fault current dissipated through their grounding system into the surrounding earth. The previously presented methodology is the only one so far known which allows evaluation and testing of the grounding system of an HV/MV substations located in urban and suburban areas by taking into account the inductive influence of the metal installations surrounding its feeding line.

Experimental investigations of the influence of the metal installations typical for urban areas on the value of the feeding line reduction factor were performed on a cable line, which in series supplies two substations in the 110-kV distribution network of Beograd, Serbia. The length of the considered line to the nearer of the supplied substations measured from the supply substation is 2.32 km, while the total length of this line is 6.59 km. The line has been made of XHLP (insulation) cables laid in a trefoil formation on the whole line length. The cross-bounding technique for circulating currents elimination was not applied to this line.

In the areas through which the line is passing, the specific soil resistivity is, on the basis of the main geological characteristics of the involved area, estimated to be in a range from 30 to 50 Ωm. The line section between the supply and the transit substation passes through an area with a lower degree of urbanization than the rest of the line. The phase conductors are made of aluminum with a cross section of 1000 mm^2, while the metallic sheaths are made of copper strings with a total cross section of 95 mm^2 and a mean diameter equal to 91 mm. The main elements of the grounding system in both of the supplied substations are the substation building foundation armature and the 44 outgoing MV cable lines made by cables with uncovered metal sheaths. The grounding impedance of the supply substation and both of the supplied substations were measured and obtained values were between 0.02 and 0.03 Ω.

The first test circuit was formed by using only the section between the supply and the transit substation. The second test circuit was formed by using the whole considered cable line. During the performed measurements in both of these cases, only the metal sheath of the cable with the test current was grounded at both of the line ends as shown in Fig. 3.13.

To obtain a complete insight into the influence of the metal installations surrounding the considered line, the currents induced in the cable sheaths were first investigated. At the beginning, this current was measured when only the sheath of the cable with the simulated ground-fault current was connected with the grounding electrodes at both ends of the line. After that one more cable sheath was connected to both grounding electrodes at the line ends, and finally, the normal operation conditions, when all three of the cable sheaths were grounded at both line ends, were explored. The measurement results show that a successive increase of the number of the grounded sheaths reduces not only the current through the grounding system of the supplied substation but also the relative participation of each of the previously connected sheaths in the ground-fault current distribution. Presented respectively, these relative reductions in the case of the sheath belonging to the cable with simulated ground-fault current are from 60.66% to 46.31% and from 46.31% to 37.49%. The result obtained for normal operating conditions when all three metal

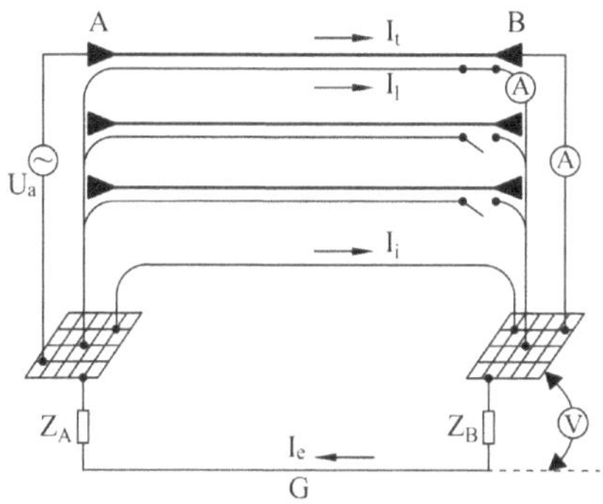

Figure 3.13 Measurement circuit that was used for the experimental analysis.

sheaths were grounded at the ends of the line were particularly important. It was showed that much smaller currents pass through the sheaths than those obtained by using the expression (2.16) in Chapter 2, Calculated reduction factor of feeding cable lines, which was derived by completely disregarding the surrounding metal installations. This result pointed out the role of the surrounding metallic installations in the ground-fault current distribution, that is, to the fact that these installations act as an additional neutral conductor of power lines passing through urban areas. According to this, they can be modeled by only one insolated or uncovered conductor laid in parallel to the considered line. This specific neutral conductor, certainly, additionally reduces the fraction of the ground-fault current dissipated into the ground through the grounding system of the supplied substation and each of the currents through the sheaths of the considered line.

By using the procedure presented in Section 3.2.2, the actual reduction factor r_a obtained for both of the considered lines was

- $r_a = -0.0314 - j\,0.0368$, or $|r_a| \approx 0.048$, for the shorter (length: 2.32 km), and
- $r_a = -0.0262 - j\,0.0228$, or $|r_a| \approx 0.034$, for the longer line (length: 6590 m).

  As can be seen, the effective values of the actual reduction factor of these lines are 53.9% and 66.9%, respectively, lower in comparison with the values obtained by using the corresponding analytical expression (2.16), by which it is not possible to take into account the influence of surrounding metal installations ($|r| \approx 0.105$).

  Then, if one assumes that the cables in the considered cases are laid in a flat formation and at a distance of $d = 0.5$ m (Fig. 3.2), under the assumption that the influence of surrounding metal installations remains unchanged, one obtains that the values are the following:

  - $r_a = -0.0352 - j0.0509$, or $|r_a| = 0.0519$ for the shorter (length: 2.32 km) and
  - $r_a = -0.0277 - j0.340$, or $|r_a| = 0.0349$, for the longer line (6590 m).

The actual values of the reduction factor, in this case, are lower by 49.7% and 64.4% than the value obtained by using the corresponding analytical expression (2.16), respectively.

In all these cases the differences are still greater in comparison with the reduction factor values obtained only based on the measurement of currents induced in the cable metal sheaths, what was the wrong, but usual practice in the past for determining the actual reduction factor of power cable lines.

Thus it is obvious that disregarding the surrounding metal installations by using the analytical expressions for the determination of the line reduction factor or determining the reduction factor only based on measurement results of currents induced in the cable sheaths gives the values which are excessively conservative. Also, it is important to note that the deviation from the actual values is greater if the cable line passes through the urban and/or suburban areas that are with a greater degree of urbanization.

The here-presented methodology enables the determination of the actual current dissipated into the surrounding earth through the grounding system of HV substations located in urban or suburban areas. Also, it enables the determination of the actual value of ground-fault current fractions passing through feeding line neutral conductors and through the metal installations surrounding the considered power line.

The essentially reduced value of the feeding line reduction factor simultaneously means a reduction by the same ratio of all potentials and potential differences (step and touch voltages) appearing on the grounding systems of supplied HV substations. Thus one can conclude that the results of the here-presented quantitative analysis throw completely new light on the grounding problem of HV substations located in urban surroundings and dramatically change our former perception of its magnitude. The actual reduction factor enables us to put the whole grounding problem of the supplied substation into the realistic frameworks that in the general case leads to the more economical solutions in each specific case. Certainly, these effects are more pronounced in cases where, because of high local soil resistivity and/or a high short-circuit current level, special measures were so far considered necessary to protect personnel and the public against too high touch voltages (e.g., bare copper wire laid in the same trench as the feeding cable line or/and in the same trench as the outgoing MV cable lines, counterpoise), for example [16,17]. Besides, one can expect the elimination of the strict requirement for the application of expensive MV cables acting as grounding conductors (cables with an uncovered metal sheath). For example, this was the case for a long time with the distribution network of Beograd.

Having in mind the similarity of urban conditions throughout the world and that the experimental investigations were performed in areas that can be considered average/ordinary from the standpoint of the achieved degree of urbanization [1,2], the previous conclusion can be considered general for the safety conditions of all distribution substations located in urban areas.

As the only inconvenience in comparison with the previous practice can be considered the fact that the actual reduction factor of each HV or EHV line depends on many local factors like they are specific constructive characteristics and spatial disposition of the metal installations surrounding the considered feeding line. It practically means that for obtaining the actual feeding line reduction factor, each distribution line passing through urban and/or suburban areas should be considered individually.

Since the power cable lines are, almost without any exception, applied in urban areas and since the favorable effects of the surrounding metal installations are very pronounced, the presented methodology could be served as a basis for the useful amendment of the current version of the corresponding technical standards, for example [13].

3.7 Methodology application at the design stage

The previously presented methodology enables compensation of the objective deficiency of all relevant, but data unknown regarding the metal installations surrounding an HV or EHV line. This is achieved, inter alia, by the test measurements of their cumulative influence upon the value of the test current in one freely chosen phase conductor and the current induced in one of the other two phase conductors used for the necessary measurements.

Thus the application of the presented methodology implies an already existing HV line. However, data regarding the actual reduction factor of an HV or EHV line, or the actual ground-fault current distribution in a supplied HV distribution substation, are usually necessary already at the design stage of these substations and their feeding lines. The problem is overcome by using the fact that, according to the complete equivalent circuit in Fig. 3.3, the current in each of the feeding line conductors contains the cumulative inductive influence of all surrounding metal installations. By taking into consideration this fact, a completely constructed line is not indispensable for obtaining the influence of surrounding metal installations. Instead of that it is sufficient only one sufficiently long cable with two conductors. The necessary field measurements would be enabled by using a provisory cable line to be laid on the surface of the soil along the foreseen route of the planned HV or EHV feeding line. For this purpose, it can be used an LV cable with two conductors sufficiently flexible for different practically possible urban situations. Also, for simulating ground-fault conditions and obtaining a corresponding electrical circuit with currents I_t and I_1 (Fig. 3.3), an auxiliary voltage source is necessary. Besides, to obtain the necessary low value of grounding impedance, the here-foreseen measurements need to be done at a stage when the grounding system of the supplied station is almost already formed.

Since the surrounding metal installations are permanently attended and have an unchangeable spatial disposition, the complete electrical circuit formed by using the provisory cable line is practically the same as that presented by the equivalent circuits shown in Fig. 3.3. The only difference stems from the fact that in this case, we have only one neutral conductor. By taking into account this difference, the voltage U_a, currents I_t and I_1, and all induced currents, according to the complete equivalent circuit shown in Fig. 3.3, are connected by the following system of equations:

$$\begin{aligned} &1.\quad Z_0 I_t + Z_{01} I_1 + \sum_{n=2}^{n=N} Z_{0n} I_n = U_a \\ &2.\quad Z_1 I_1 + Z_{01} I_t + \sum_{n=0}^{n=N} Z_{1n} I_n = 0, \quad n \neq 1 \end{aligned} \tag{3.36}$$

According to Eq. (3.36), the values of currents I_t and I_1 contain the cumulative effect of all, known and unknown, surrounding metal installations. Thus when currents I_t and I_1 are obtained by the corresponding measurements, all surrounding metal installations can be substituted by only one equivalent neutral conductor. Then, the relevant parameters of this conductor and the actual reduction factor of the designed HV or EHV line can be determined by using the calculation procedure analogous to the one presented in Section 3.2.2.

Since metal installations in urban areas are placed mainly under the soil surface, the obtained influence of the surrounding metal installations will be slightly smaller in the case of a planned cable (underground) line or slightly greater in the case of a planned overhead line. Since the overhead line conductors are placed at a greater distance from the surrounding metal installations, the value of the radius of the equivalent cylindrical conductor representing the surrounding metal installations in the case of these lines should be enlarged by the average pole height. Because of this approximation, the accuracy of the presented methodology is somewhat smaller at the design stage of future HV distribution lines.

3.8 General observations concerning the methodology application

Although there are several methods of measuring soil resistivity, none of them can give sufficiently trusted results when applied in urban conditions. The reason for that is in the fact that surfaces of urban areas are already covered by buildings, streets, sidewalks, and many other permanently constructed objects, while below-ground many known and unknown metal installations already exist. Thus we are forced to adopt an approximate value of the equivalent soil resistivity, based on the main geological characteristics of the surrounding area, and use it in the calculations necessary for the application of this methodology. Here, the favorable circumstance is that the self- and mutual impedances represent logarithm functions of the equivalent soil resistivity and only slightly dependent on it. Because of that approximate data concerning equivalent soil resistivity have no relevant influence on the accuracy of the finally obtained results. It is sufficient to know that, within the framework of possible values of equivalent soil resistivity, the lowest one should be preferred because it gives final results that are slightly on the side of larger security. In the case of an existing HV line when mutual distances between the phase conductors are significant, the here-defined measurements should be performed for each of the phase conductors to obtain the largest (critical) value of the actual reduction factor.

Having in mind that the here-presented analytical procedure is developed based on the results of the on-site measurements, the presented methodology takes into account the influence of all relevant factors and parameters, including those influence of which is negligibly small. Thus the accuracy of the presented methodology depends mainly on the accuracy of the measurements of the simulated ground-fault current in one adopted phase conductor and current induced in one of the two other phase conductors, which, in principle, can be freely chosen. This means that the presented methodology enables the determination of the actual reduction factor for any line existing in practice independently of its constructive characteristics, the complexity of mutual inductive influences of different surrounding metal installations, and the number of relevant but unknown data.

Some inaccuracies can also appear as a consequence of the omnipresent power frequency, 50 or 60 Hz. This influence can be efficiently avoided by using a test current of somewhat higher frequency that can easily be discriminated from the omnipresent power frequency. The introduced inaccuracy is small and gives final results that are slightly on the side of enlarged safety. Also, for the application of the presented methodology, it is important to note that the simulated ground-fault current should circulate along the whole feeding line length. Only in that way we can be sure that we take into account all relevant data concerning the ground-fault current distribution in the supplied substation.

Since necessary test measurements are performed with a simulated ground-fault current significantly smaller compared to the actual ones, the following should also be mentioned the results of several experimental measurements of the reduction factor performed by experimentally obtained real ground-fault conditions in the 110-kV distribution network of Beograd. They showed good agreement with the results obtained using almost 100 times smaller simulated earth fault currents. Finally, all foreseen test measurements and corresponding calculations are relatively simple and easily feasible in each, practically possible case.

It is quite realistic to assume that the lines consisting of the same cables and belonging to the same urban area have similar values of the reduction factor. Because of that it can be useful to know the actual reduction factor of the existing HV or EHV lines in a certain power distribution network. The values of the reduction factors of these lines could be valuable information to the designers of future HV/MV substations and their feeding lines. Based on these values, the designers of future HV/MV substations and their feeding lines can obtain a necessary engineer feeling concerning the order of the magnitude of the actual reduction factor of HV and EHV lines in certain urban areas.

References

[1] Lj.M. Popović, Influence of the metallic installations surrounding the feeding cable line on the ground fault current distribution in supplied substations, IEEE Transactions on Power Delivery 23 (4) (2008) 2583–2590.

[2] Lj.M. Popović, Testing and evaluating grounding systems for substations located in urban areas, IET Generation Transmission and Distribution 5 (2) (2011) 231–238.

[3] Lj.M. Popović, Determination of the actual reduction factor of HV and MV cable line passing through urban and suburban area, Facta Universitatis, Series: Electronics and Energetics 27 (1) (2014) 25–39.

[4] Lj.M. Popović, Ground-fault current distribution when a ground fault occurs in HV substations located in urban areas, Progress in Electromagnetic Research B 5 (2014) 167–179.

[5] Lj.M. Popović, Reduction of the fault current passing through the grounding system of an HV substation by cable line, International Journal of Electrical Power & Energy Systems 99 (0) (2018) 493–499.

[6] Lj.M. Popović, New method for determination of inductive influence of one or more power lines on any exposed metal installations, International Journal of Engineering Research and Applications 6 (2) (2016) 36–46.

[7] Lj.M. Popović, Inductive influence of HV cable lines in urban and suburban areas, Electrical Power System Research 176 (2019). Available from: https://doi.org/10.1016/j.epsr.2019.105944.

[8] Lj.M. Popović, Influence of surrounding metal installations on the parameters of electric-power lines passing through urban and suburban areas, in: Materials and Processes for Energy: Communicating Current Research and Technological Developments, Formatex Research Centre, Energy Book Series – 2013 Edition, Spain, 2013, pp. 831–839.
[9] Lj.M. Popović, Transfer characteristics of electric-power lines passing through urban and suburban areas, International Journal of Electrical Power & Energy Systems 56 (2014) 151–158.
[10] Lj.M. Popović, Inductive influence of surrounding metal installations on power transfer capacity of HV cable lines with applied cross-bonding, International Journal of Electrical Power & Energy Systems 120C (2020) 105963. Available from: https://doi.org/10.1016/j.ijepes.2020.105963.
[11] Lj.M. Popović, Preliminary testing and assessment of safety conditions of HV substations located in urban areas, Electric Power System Research 195 (2021). Available from: https://doi.org/10.1016/j.epsr.2021.107111.
[12] Lj.M. Popović, Practical Method for Analysis and Design of HV Installation Grounding Systems, Elsevier, London, 2018.
[13] Short-Circuits Currents in Three-Phase A.C. System-Part 3: Currents During Two Separate Simultaneous Line-to-Earth Short Circuit and Partial Short-Circuit Flowing Through Earth, IEC Int. Std. 60909-3, (Ed.) II, 2003.
[14] Lj.M. Popović, Comparative analysis of grounding systems formed by MV cable lines with either uninsulated or insulated metal sheath(s), Electric Power System Research 81 (2) (2011) 393–399.
[15] IEEE Guide for Bonding Shields and Sheaths of Single-Conductor Power Cables Rated 5 kV through 500 kV, IEEE Standard 575-2014, 2014.
[16] S.T. Sobral, V.S. Costa, M.S. Campas, D. Mudhedkar, Dimensioning of nearby substations interconnected grounding systems, IEEE Transactions on Power Delivery 5 (3) (1988) 1605–1614.
[17] A. Campuccia, G. Zizzo, A study of the use bare buried conductors in an extended interconnected earthing system inside an MV network, in: Proc. CIRED Conference, Turin, 2005.

Further reading

IEEE Guide for Safety in AC Substation Grounding, 2000 IEEE Guide for Safety in AC Substation Grounding, IEEE Std. 80, 2000.
Kersting, 2002 W.H. Kersting, Distribution System Modeling and Analysis, first ed., CRC Press LLC, 2002, pp. 95–105.
Popović, 2003 Lj.M. Popović, Determination of the reduction factor for feeding cable lines consisting of three single-core cables, IEEE Transactions on Power Delivery 18 (3) (2003) 736–743.
Popović, 2008 Lj.M. Popović, Improved analytical expressions for the determination of the reduction factor of the feeding line consisting of tree single-core cables, European Transactions on Electrical Power 18 (2) (2008) 190–203.

Reduction factor of lines consisting of one overhead and one cable section

4

4.1 Introduction

In an high-voltage (HV) or extra-HV distribution network, during a phase-to-ground fault, many current flows and raised potentials appear at places where they do not exist under normal operating conditions. The magnitudes of these potentials, as well as all of the potential differences relevant for the evaluation of safety conditions (touch and step voltages), are proportional to the current dissipating through the substation grounding system into the surrounding earth.

The grounding system of an HV substation, in general, justifies its existence only at the moment of an unbalanced fault when a high fault current leaves the power system and increased potentials appear at places exposed to human touch. To ensure protection against undesired consequences, such as the loss of human lives, burn down of grounding wires, and damages to sensitive electronic equipment that will be used in the future substation, it is necessary to design and construct the grounding system satisfying the strictly prescribed safety criteria. During the design procedure, it is necessary to estimate, as accurately as possible, the values of potentially dangerous step and touch voltages at any place within and in the vicinity of the designed substation. When the estimated maximal values of these voltages are within the tolerable limits, the designed grounding system is correctly designed. However, the designer is also asked to make the solution as cheaper as possible.

The magnitudes of all potentials as well as of all potential differences, including touch and step voltages, are proportional to the current dissipated through the substation grounding system into the surrounding earth. Since the ground-fault current in HV networks with the directly grounded neutral points is usually very high, the favorable circumstance is that this current is only a fraction of the total ground-fault current. The rest of the ground-fault current uses all other available return paths like transmission line ground wires, cable sheaths, and other surrounding metal installations grounded at both ends. Thus these parts of the ground-fault current will not participate in the creation of any potential in the grounding system of a designed substation. Therefore for the designer of a substation grounding system, it is of prime importance to have on a disposal a method enabling as much as possible accurate the determination of the ground-fault current distribution. Since the substation grounding electrode/grid is usually the only element of the whole grounding system that is designed and constructed exclusively for grounding purposes, it is especially important to know the part of the earth current dissipated into

Relevant Characteristics of Power Lines Passing through Urban Areas. DOI: https://doi.org/10.1016/B978-0-323-91136-8.00004-9

the ground via this element of the grounding system. Thus only after correctly determined this part of the ground-fault current, it is possible to see the whole problem into a realistic framework and to arrive at safe solutions without excessive expenditures.

Evaluation of the ground-fault current distribution for faults in HV substations supplied by overhead (transmission) lines that can be treated as lines with uniformly distributed parameters along their entire length is possible by using relatively simple analytical expressions. These expressions for various substations and for different types of transmission lines are presented in Ref. [1]. Also, in the case of cable lines that have been constructed as lines with uniformly distributed parameters along their entire length, adequate analytical expressions are developed [2–5] and presented here in Chapter 2, Calculated reduction factor of feeding cable lines.

However, the mentioned analytical expressions are not relevant to all cases that can be encountered in contemporary power distribution practice and cannot be used in the cases where a feeding line has two or more sections with one or more mutually different parameters relevant for the ground-fault current distribution. Certainly, in such cases, the problem gains in complexity and deserved a separate consideration. With the aim to efficiently solve the problem in these relatively rare cases, several methods have been so far developed, for example [6–13].

The possibility of simplifying the problem and obtaining an analytical solution has been seen in the reduction of the number of the elements of the electrical circuit spontaneously formed via the different neutral conductors of this type of feeding line during a ground fault. To this purpose the formula that enables substituting lumped parameter ladder circuits of any size (from one to an arbitrarily large number of π-cell) by only one equivalent π-cell was used [14–16]. Thus the development of the method is based on the mentioned solution for lumped parameter ladder circuits, and the assumptions and approximations that are anyway unavoidable at the design stage. The finally derived analytical expressions enable the estimation of the ground-fault current distribution when the feeding line is composed as a certain combination of one cable and one overhead section.

Large generating substations supplying HV distribution networks of big cities require too large unoccupied plots and as a result are built outside of urban zones. Outgoing lines from these substations supply HV distribution substations (e.g., 110/10 kV), situated deep within urban areas, that is, in the center of electric energy consumption. Each of the outgoing lines usually supplies two or three such substations in succession, forming the so-called radial distribution system. If these lines are made of cables along the whole of their lengths, from the source substation to the supplied HV substation, calculation of ground-fault current distribution for a fault in the supplied substation is considered in Chapter 2, Calculated reduction factor of feeding cable lines. However, since these lines usually pass through areas of different degrees of urbanization, one sometimes has a situation where a certain substation is supplied by a cable line that is a continuation of an overhead line [6,8,11,12]. With a permanent elevation of the ground-fault current level and with an increasing deficiency of free and sufficiently large urban space for the building of new HV substations, correct determination of ground-fault current distribution in various practically possible conditions and circumstances certainly gains more and more in importance.

A situation when a feeding line is composed of one cable and one overhead section is symbolically depicted in Fig. 4.1.

According to the situation depicted in Fig. 4.1, the problem of ground-fault current distribution becomes somewhat more complicated in such cases. The current that appears due to the inductive coupling between faulty phase conductors on one side and ground wire(s) and cable metal sheath(s) on the other side cannot be determined by the analytical expressions presented in Chapter 2, Calculated reduction factor of feeding cable lines. One part of the current induced in cable sheath(s), at the place of discontinuity, continues its flow toward the supply sources through the ground surrounding the feeding line. As a result, the reduction in fault current through the grounding system of the supplied substation is somewhat smaller than in the case where the feeding line is made of cables along the whole of its length. On the other hand, this reduction is larger than in the case where the feeding line is along its whole length overhead. In earlier practice, however, when determining the fault current distribution, such a line is sometimes treated either as only a cable or as only an overhead line. In the first case, this may result in a serious error at estimating the achieved safety conditions within and in the vicinity of the designed HV substation, that is, they can be considered satisfied, although this is not the case.

Because of the strong inductive coupling between the faulty phase conductors and cable sheath(s), most of the ground-fault current can be injected into the ground at the place of discontinuity rather than at the fault place, supplied substation *B*. When the place of discontinuity is in a transit substation, this effect is described as the "fault application transfer effect," because a large part of the ground-fault current is transferred to the transit substation and dissipated into the earth via the grounding system of this substation [6,12].

Examples of such feeding lines are encountered in distribution networks with various voltage levels and as already mentioned are a consequence of various degrees of urbanization of the areas through which a feeding line is passing. In current power distribution practice, such cases also appear where a designed substation should be located nearby the existing overhead line. Then we have a situation that a foreseen new substation *B* should be connected by interpolation at point *D* to the

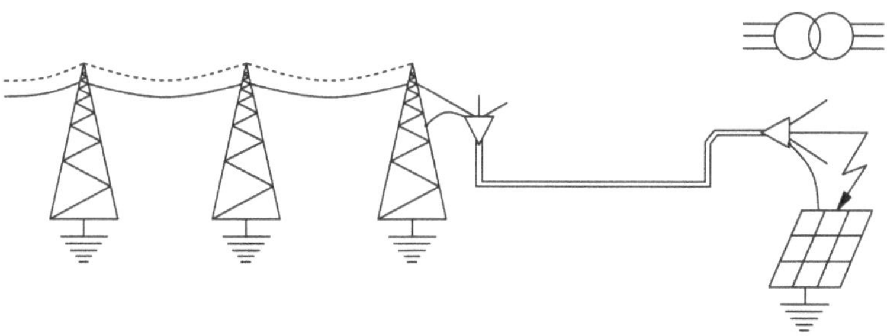

Figure 4.1 Feeding line composed of one cable and one overhead section.

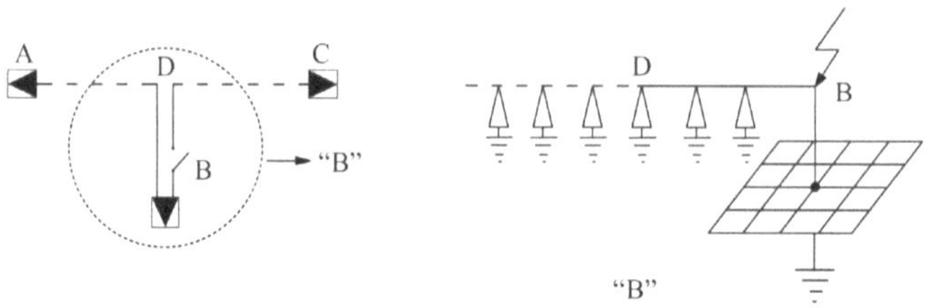

Figure 4.2 Feeding line consisting of two different overhead sections.

existing overhead line that previously connected stations *A* and *C* directly, as is symbolically shown in Fig. 4.2.

According to Fig. 4.2, the feeding line is divided into one planned and one already existing section. As a rule, the existing section is equipped with one steel ground wire, whereas the new section has two aluminum conductor steel-reinforced wires (ACSRs).

On the basis of Figs. 4.1 and 4.2, it is certain that the current induced in the neutral conductor(s) of the line section nearer to the supplied substation is larger than the current induced in the neutral conductor(s) of the rest of the line. Therefore one part of the current induced in the section nearer to the supplied substation at the point of discontinuity has to continue its path toward its sources through the earth.

4.2 Calculated reduction factor

4.2.1 Cable having uncovered metal sheath

In an HV network with directly grounded neutral points, the ground-fault current distribution in phase conductors and the ground-fault current distribution in its return paths represent two mutually independent distributions and because of that can be considered separately. Bearing in mind this fact, we will assume here that the ground-fault current passing through the fault place is known from the prior ground fault studies and as such will not be a subject of our consideration.

Now, let us assume that the ground fault occurs within a distribution substation *B* supplied from substation *A* by a cable section of the feeding line having an overhead section, as well. Then, we will assume from the standpoint of the line equivalent circuit during a ground fault the most complex case when the foreseen cable has an uncovered metal sheath and the overhead section has a ground wire grounded at each pillar. Under such assumption by using the following theoretical bases:

- method of symmetrical components,
- driving point technique,
- decoupled method, for example [6], and
- the analytical expressions for reduction of an arbitrarily large ladder circuit to only one equivalent π-cell [14–16],

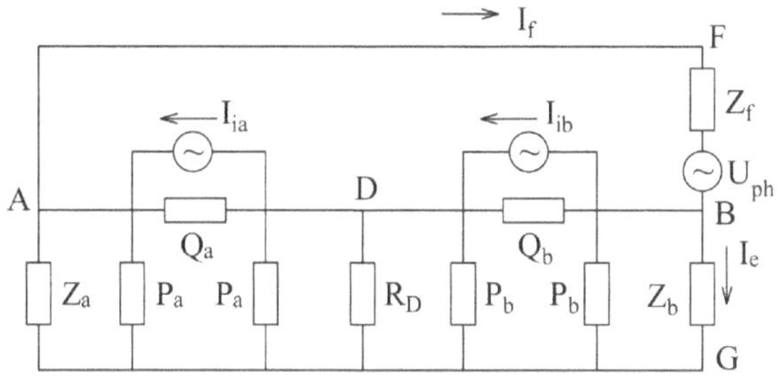

Figure 4.3 The equivalent circuit of the assumed line during a ground fault.

we obtain the equivalent circuit shown in Fig. 4.3.

The used notation has the following meaning: F is the fault place (substation B); U_{ph} is the auxiliary driving voltage source at fault place with the voltage equal to the phase voltage at the source station before the fault; I_f is the ground-fault current; I_e is the current dissipated into the earth through the ground grid of substation B and the connected external grounding electrodes (in general, they can be metal sheaths of connected cable lines, neutrals of the low voltage network, metal pipelines, etc.) excluding the grounding effects of the considered feeding line, I_{ia} is the current source that replaces the influence of the inductive coupling between the faulty phase conductor and the ground wire(s); I_{ib} is the current source that replaces the influence of the inductive coupling between the faulty phase conductor and the metal cable sheath(s); Z_f is the impedance of the electric power system at the fault place; Q_a and P_a are the impedances that make up the equivalent π-cell for the ladder network formed by the ground wire; Q_b and P_b are the impedances that make up the equivalent π-cell representing the cable sheath as a grounding conductor; $Z_a(Z_b)$ is the impedance of the grounding system of substation A (B) which does not include the grounding effects of the considered line; R_D is the footing resistance of the tower at the discontinuity place; D is the place of line discontinuity; and G is the remote ground.

When substation B has a grounded neutral point(s), the equivalent circuit from Fig. 4.3 should be completed with a branch containing the zero-sequence impedance of the local transformer and connects points F and B.

As can be seen in reality very complex electrical circuit with numerous mutually conductively and inductively coupled elements is presented in Fig. 4.3 which is significantly simplified by reducing the number of the elements of lumped parameter ladder circuits. The elements representing the cable section as a grounding conductor, impedances Q_b, and P_b, are determined, according to Refs. [15,16], by

$$Q_b = \frac{k^{L_c} - k^{-L_c}}{2} Z_{c\infty}, \tag{4.1}$$

$$P_b = \frac{k^{L_c} + 1}{k^{L_c} - 1} Z_{c\infty}, \tag{4.2}$$

where k represents the coefficient of current distribution at each cable segment of 1-m along an infinitely long cable line, determined by

$$k = 1 + \frac{Z_{c\infty}}{R_{sh}}, \tag{4.3}$$

and where $Z_{c\infty}$ is the grounding impedance of an infinitely long cable line, determined by

$$Z_{c\infty} = \frac{Z_{sh}}{2} + \sqrt{R_{sh}Z_{sh} + \frac{Z_{sh}^{2}}{4}} \approx \sqrt{R_{sh}Z_{sh}}, \tag{4.4}$$

whereas L_c represents the relative cable section length expressed in relation to the length of 1 m.

Variables R_c and Z_{sh} in Eq. (4.4) denote discrete parameters of the cable segment of 1 m. Thus resistance R_c is, according to, for example [16], determined by

$$R_{sh} = \frac{\rho}{\pi m} \ln \frac{L_c}{\sqrt{d_{sh}h}}, \quad \Omega, \tag{4.5}$$

where ρ is the equivalent soil resistivity, d_{sh} is the outer diameter of the cable sheath (m), h is the cable burial depth (m), and m is the length of 1 m,

whereas impedance Z_{sh} is determined by

$$Z_{sh} = Z'_{sh}(\Omega/m) \times 1 \text{ m}, \tag{4.6}$$

where Z'_{sh} is the self-impedance of the cable sheath determined by Eq. (4.7).

The current induced in the metal sheath(s) of the cable section is, according to the reduction factor definition in Chapter 2, Calculated reduction factor of feeding cable lines, determined by the following expression:

$$I_{ib} = (1 - r)I_f, \tag{4.7}$$

where r is the reduction factor of the line under the assumption that the cable(s) are laid along the whole line length (for the three-core cables usually given as factory data).

The analytical expressions for the reduction factor for different types of cable lines and different cable formations are presented in Chapter 2, Calculated reduction factor of feeding cable lines.

For the part of the equivalent circuit representing the line overhead section with an arbitrary number of spans, N, the impedances Q_a and P_a, are, according to Refs. [14–16], determined by

$$Q_a = \frac{k^N - k^{-N}}{1 + k} Z_\infty, \tag{4.8}$$

$$P_a = \frac{k^N - k^{-N}}{k^N + k^{-N+1} - k - 1} Z_\infty, \tag{4.9}$$

where k represents the coefficient of current distribution at each tower along an infinitely long line.

$$k = 1 + \frac{Z_\infty}{R}. \tag{4.10}$$

In expression (4.10), Z_∞ designates the grounding impedance of the ground wire under the assumption that the line is, from the standpoint of grounding effects, infinitely long. This impedance is determined by

$$Z_\infty = \frac{Z_s}{2} + \sqrt{RZ_s + \frac{Z_s^2}{4}}. \tag{4.11}$$

where R is the average tower footing resistance, and Z_S is the self-impedance of the ground wire on one overhead section span of the average length.

The impedance Z_S is determined by

$$Z_S = L_S Z'_{gw}, \quad \Omega, \tag{4.12}$$

where L_S is the average length of the spans overhead section of the feeding line (km), and Z'_{gw} is the self-impedance of the ground wire per unit length (Ω/km).

The self-impedance of the ground wire is determined by

$$Z'_{gw} = R'_{gw} + \omega\frac{\mu_0}{8} + j\omega\frac{\mu_0}{2}\left(\frac{\mu_r}{4} + \ln\frac{\delta}{r_{gw}}\right), \quad \Omega/\text{km} \tag{4.13}$$

where R'_{gw} is the longitudinal resistance of the ground wire (Ω/km); r_{gw} is the equivalent ground wire radius of one or more ground wires (m); ω is the angular power frequency: $2\pi f$; μ_0 is the magnetic permeability of vacuum: $4\pi \times 10^{-7}$ Vs/Am; μ_r is the relative magnetic permeability of the installation material; and δ is the equivalent earth penetration depth (m).

The current induced in the overhead section of the line, expressed through the reduction factor of an overhead line, is determined by

$$I_{ia} = (1 - r_0)I_f, \tag{4.14}$$

The reduction factor of the overhead line r_0 is, according to Refs. [1,17], determined by

$$r_0 = 1 - \frac{Z'_m}{Z'_{gw}}, \tag{4.15}$$

In expression (4.15), Z'_m represents the mutual impedance between the ground wire and the faulty phase conductor that is, according to Ref. [17], determined by

$$Z'_m = \omega\frac{\mu_0}{8} + j\omega\frac{\mu_0}{2}\ln\frac{\delta}{d_w}, \quad \Omega/\text{km}, \tag{4.16}$$

where d_w is the distance between the ground wire(s) and the phase conductor carrying ground-fault current (m),

Since to estimate the safety conditions, it is necessary to determine the largest possible value of the current I_e, impedance Z'_m should be determined under the assumption that the fault occurred on the phase conductor farthest from the ground wire.

The equivalent circuit shown in Fig. 4.1 does point out the "proximity effect between the cable as a grounding conductor and the ground grid" of substation *B*. However, in the considered case the mentioned effect has a very small influence on ground-fault current distribution because of the strong inductive coupling between the faulty phase conductor and the cable sheath(s). Due to this, when the length of a cable section is sufficiently large, most of the ground-fault current dissipates into the earth at the place of discontinuity *D*, instead of through the grounding system of substation *B* [11].

Also, in practical conditions, the influence of the impedance Z_a on the ground-fault current distribution is so small that this impedance can be neglected ($Z_a \approx 0$). Thus to determine the ground-fault current distribution, we can use the simplified equivalent circuit shown in Fig. 4.4.

In the circuit in Fig. 4.4, ΔI_i denotes the current representing the difference between induced currents I_{ib} and current I_{ia}.

According to this equivalent circuit, we can write a closed system of equations, solution of which gives the following expression for the coefficient expressing the total reduction of current I_e:

$$r_t = \frac{I_e}{I_f} = \frac{r_0(1-r)Z_D + r(Q_b + Z_D)}{\left(1 + \left(Z_b/P_b\right)\right)(Q_b + Z_D) + Z_b}. \tag{4.17}$$

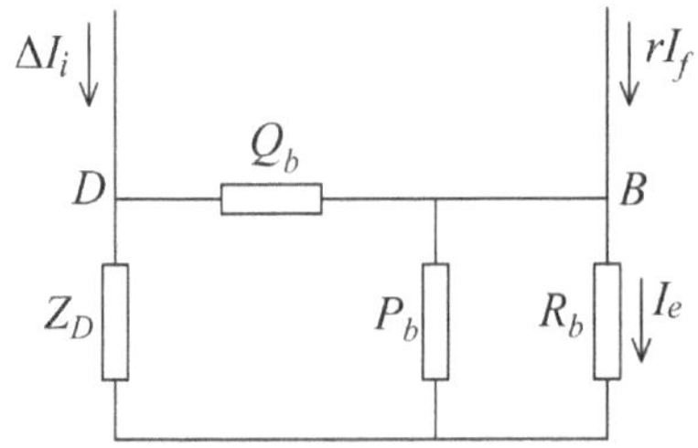

Figure 4.4 Simplified equivalent circuit.

In expression (4.17), grounding impedance Z_D is, according to Figs. 4.3 and 4.4, determined by

$$Z_D = \left(\frac{1}{R_D} + \frac{1}{Z_N} + \frac{1}{P_b}\right)^{-1}, \tag{4.18}$$

where impedance Z_N can be, under the assumption $Z_A \approx 0$ and according to Fig. 4.3, determined by the following approximation:

$$Z_N \approx Z_N(Z_A \approx 0) = \frac{k^N - k^{-N}}{k^N + k^{-N+1}} Z_\infty, \tag{4.19}$$

The coefficient r_t expresses the total influence of the considered feeding line on the ground-fault current distribution because it, besides the inductive influences of the cable sheath and ground wire, takes into account also their grounding effects in the reduction of the ground-fault current fraction that is dissipated into the earth, I_e.

Since it should express only the inductive effects of the line neutral conductors, according to the definition of the reduction factor of feeding lines, all grounding effects should be omitted from Eq. (4.19). We will eliminate them, according to the circuit in Fig. 4.4, by introducing the following approximation: $Z_b = 0$. Then, on the basis of Eq. (4.17), we obtain

$$r_i = r + \frac{r_0(1-r)Z_D}{Q_b + Z_D}. \tag{4.20}$$

Thus coefficient r_i expresses only the influence of the inductive couplings, between the phase conductor and metal sheath(s) of the cable section as well as between the same that phase conductor and ground wire(s) of the overhead section, on the distribution of the ground-fault current.

The second addend in Eq. (4.20) obviously represents the increment of the reduction factor in comparison to its value obtained under the assumption that the feeding line is performed by cable(s) along the whole of its length. If we assume that the length of the cable section infinity increases, then the value of impedance Q_b tends to infinity ($Q_b \rightarrow \infty$) and the value of the reduction factor r_i tends to r ($r_i \rightarrow r$). Also, according to Eq. (4.20) reduction factor r_i depends on the value of the grounding impedance at the place of discontinuity, Z_D. With a decrease of this impedance, if other relevant parameters remain unchanged, the value of the reduction factor r_i also decreases. If the value of this impedance becomes negligibly small, the value of the reduction factor of the feeding line becomes practically equal to the reduction factor of the cable section, or $r_i \approx r$. On the other side, with an increase of impedance Z_D, if other relevant parameters remain unchanged, the value of r_i increases and tends to the value of the reduction factor of the overhead line, r_0. Since under practical conditions, the reduction factor of a cable line is significantly lower than the reduction factor of an overhead line ($r \ll r_0$), it can be said that in each concrete case the value of r_i is situated between the two limiting values, r and r_0 ($r < r_i < r_0$).

4.2.2 Special cases

The presented equivalent circuit and the given analytical expressions are valid for the case that can be considered a general one, from the standpoint of grounding effects of neutral conductors, but can easily be modified for various special cases that can be encountered in practice.

When the length of the overhead section is such that we can consider it infinite from the point of view of the grounding effects, instead of Eq. (4.18), we can use

$$Z_D = \left(\frac{1}{R_D} + \frac{1}{Z_\infty} + \frac{1}{P_b}\right)^{-1}, \tag{4.21}$$

where Z_∞ denotes the grounding impedance of the ground wire under the assumption that the ground wire (overhead section) is infinitely long from the standpoint of grounding effects.

An overhead section with N spans can be considered infinite, from the standpoint of the grounding effects of its grounding wire, if its length is larger or equal to its active length. The active length of an overhead line (ground wire) can be, according to Refs. [14,16], determined by the following approximation:

$$N_a(\varepsilon = 0.05) \approx \frac{3}{\ln\left(1 + \sqrt{\left(\left|L_S Z'_W\right|/R\right)}\right)}, \tag{4.22}$$

where N_a is the active length of the ground wire(s), expressed through the number of line spans; ε is the relative error; R is the average tower footing resistance (Ω); and L_S is the length of one span of average length (km).

When the place of discontinuity is a transition substation, the case considered in Ref. [6], impedance Z_D is equal to the grounding system impedance of this substation, or $Z_D = Z_{gt}$, where Z_{gt} denotes the grounding system impedance of the transit substation.

In the case of a steel ground wire, the reduction factor r_0 has a large value ($r_0 \approx 0.95$) so that expressions (4.17) and (4.20) can be simplified by introducing the approximation: $r_0 \approx 1$. Also, when the overhead section is without ground wire, then we have: $r_0 = 1$ and $Z_D \approx P_b$, so that, instead of Eq. (4.20), we have

$$r_i = \frac{rQ_b + P_b}{Q_b + P_b}. \tag{4.23}$$

Finally, in the case when the length of the cable section increases indefinitely, according to Eqs. (4.1) and (4.2), Q_b tends to infinite and P_b to $Z_{c\infty}$, so that, instead of Eq. (4.23), we have

$$\begin{aligned} &\lim r_i = \lim \frac{rQ_b + P_b}{Q_b + P_b} = r \\ &Q_b \to \infty \quad Q_b \to \infty \\ &P_b \to Z_{c\infty} \quad P_b \to Z_{c\infty}. \end{aligned} \tag{4.24}$$

Based on Eq. (4.24), we have

$$r_i(L \geq L_a) = r, \tag{4.25}$$

As can be seen from Eq. (4.25), the reduction factor of a cable section longer than its active length, from the standpoint of its grounding effects, becomes equal to the reduction factor of the feeding line that is laid along its whole length as a cable line. The active cable line length from the standpoint of the grounding effects of its sheath can be, according to Ref. [18], estimated based on the following approximation:

$$L_a(\varepsilon = 0.05) \approx \frac{3m}{\ln\left(1 + \sqrt{\left(\pi |Z_{sh}| m / \rho \ln\left(L_a / \sqrt{h d_{sh}}\right)\right)}\right)}, \tag{4.26}$$

where L_a is the active cable line length in meters (m); ε is the relative error; ρ is the equivalent soil resistivity; d_{sh} is the outer diameter of the cable sheath (m); h is the cable burial depth (m); and m is the length of 1 m.

Based on transcendental Eq. (4.26), the active length of cable lines with an uncovered metal sheath can be estimated for each such cable line.

4.2.3 Cable(s) having covered metal sheaths

This type of cable line does not have grounding properties, that is, do not dissipate the ground-fault current through its metal sheath into the surrounding earth, because their metal sheath is covered by an insulating sheath. Based on this fact, it is necessary to do a certain modification of the equivalent circuit shown in Fig. 4.4. Impedance P_b should be omitted ($P_b = \infty$), whereas impedances Q_b and Z_D, according to Fig. 4.3, become modified into the following:

$$Q_b = LZ'_c, \tag{4.27}$$

$$Z_D = \left(\frac{1}{R_D} + \frac{1}{Z_N}\right)^{-1}, \tag{4.28}$$

where L is the length of the cable section (km), and Z'_{sh} is the self-impedance of the cable sheath(s) per unit length (Ω/km).

Then, on the basis of Eqs. (4.20) and (4.27), we obtain

$$r_i = r + \frac{r_0(1-r)Z_D}{LZ'_c + Z_D}. \tag{4.29}$$

When the overhead section regarding the grounding effects can be considered infinitely long, instead of Z_N, we have Z_∞ in Eq. (4.28) and when a place of discontinuity is a transition station, then we have $Z_D = Z_{gt}$. This case can be encountered in the HV distribution networks. The actual values of L and Z_{gt} in these networks are such that the condition $Z_{gt}/LZ_{sh} < 0.05$ is satisfied in almost all cases and instead of Eq. (4.26) can be used the following simple approximation: $r_i \approx r$.

If the discontinuity place is on the line itself, the reduction effect of this type of cables lines is less pronounced because the grounding impedance Z_D, in that case, is significantly larger ($P_b = \infty$). Thus in some cases, it could be more economical to undertake the measures leading to a lower value of R_D Eq. (4.18) than the measures leading to a lower value of the grounding resistance of the supplied substation ground grid. Certainly, a final decision concerning this can be made only after the comprehensive consideration conducted in each concrete case.

In the medium-voltage (MV) distribution networks, it is common that several substations are successively supplied through the same cable section. This situation is symbolically depicted in Fig. 4.5.

However, since the lumped parameter ladder circuit formed by the cable sheath (s) and the grounding electrodes of the supplied substations is analogous to the ladder circuit formed by the overhead ground wire, this network can also be represented by only one equivalent π-cell. In that case, elements Q_b and P_b can also be determined by Eqs. (4.8) and (4.9). It is necessary only at the determination Q_b and P_b instead of Z_S and R, to use the average value of the self-impedance of the cable sheath(s) between the two adjacent substations and the average value of the grounding resistance of the substations supplied directly by the cable section of the feeding line. The ground-fault current distribution when a ground fault occurs in substation B is symbolically depicted by Fig. 4.6.

The used notation has the following meaning: A, B, and C are the substations supplied by the cable section of the feeding line; I_f is the ground-fault current; I_i is the current appearing in the cable sheath solely as a consequence of the inductive coupling between the cable sheath(s) and the faulty phase conductor; I'_g is the current dissipated into the earth through the grounding system of substation B; I_g is the current dissipated into the surrounding earth through the ground electrode/grid of substation B; I_{gA} (I_{gC}) is the current appearing in the cable sheath(s) solely as a consequence of the conductive connection with the grounding system of substation A (C); Z_a (Z_b) is the grounding system impedances of the substation A (C) without the

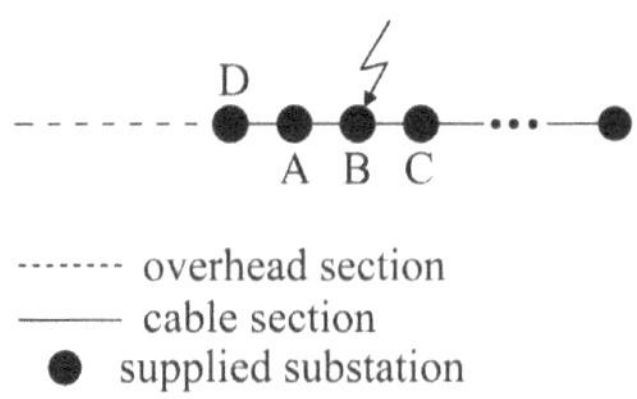

Figure 4.5 Cable section supplying several MV/LV substations.; *LV*, Low voltage; *MV*, medium voltage.

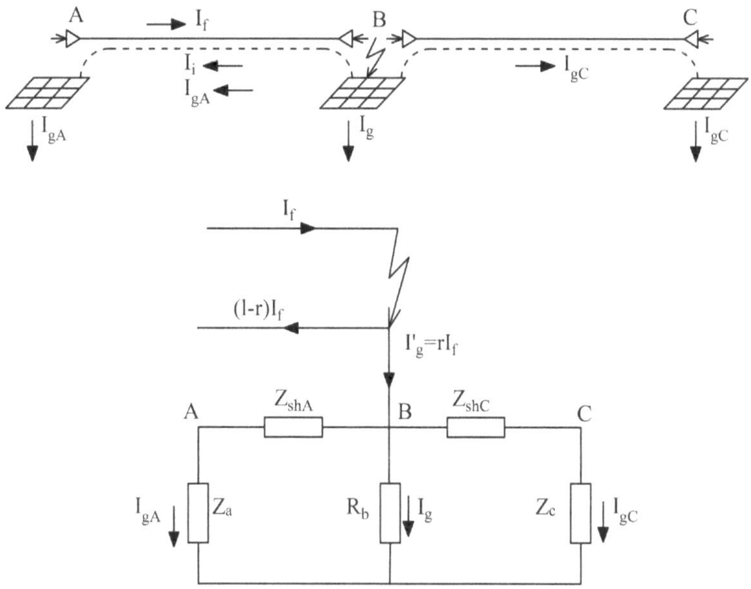

Figure 4.6 Ground-fault current distribution.

grounding effects of cable section BA (BC); Z_{shA} (Z_{shC}) is the self-impedance of the cable sheath(s) from substation A (C) to B; R_b is the resistance of the grounding electrode (grid) of substation B; and r is the reduction factor of the feeding line determined by Eq. (4.20).

The presented analysis showed that the reduction of current I_g in the case of a feeding line consisting of a cable and an overhead section depends on factors such as the cable section length, the type of the used cable (with an insulated or with an uncoated outer sheath), the grounding impedance at the point of discontinuity, as well as reduction factors relevant for individual line sections r and r_0.

Based on Eq. (4.29), it is important to note that the reduction factor for a sufficiently long cable section becomes equal to the reduction factor of the feeding line that is the cable line on the whole of its length. The same is true when, independently of the cable section length, the grounding impedance at the place of discontinuity Z_D is negligibly small ($Z_D \approx 0$).

By using here developed analytical expressions, it is possible to determine the calculated reduction factor of any distribution line composed of two mutually different sections. However, by using these expressions one cannot take into account the favorable inductive influence of the surrounding metal installations typical for urban areas. The inductive influence of these installations on the ground-fault current distribution in substations located in an urban area is so intensive that the actual reduction factor of their feeding lines is around three times lower than the one obtained by calculations, as shown in Chapter 3, Actual reduction factor of power distribution lines.

4.3 Actual reduction factor

The analytical expressions presented in Chapter 2, Calculated reduction factor of feeding cable lines, enable taking into consideration the relevant constructive characteristics of the cables as well as the formation in which single-core cables can be laid. Besides, the analytical expressions presented here enable taking into account an even somewhat larger number of relevant data relating to the line composed of one cable and one overhead section. However, none of them enables the determination of ground-fault current distribution in substations located in urban and suburban areas, where many other metal installations participate in the ground-fault current returning to the power system.

The actual reduction factor involves the inductive influence of all known and unknown metal installations surrounding a feeding line and because of that depends on many local conditions and circumstances and cannot be determined only by the previously presented analytical expressions. This reduction factor of the lines passing through urban and/or suburban areas can be determined in all cases, including the cases considered in this chapter, by using the methodology based on the testing measurements described in Chapter 3, Actual reduction factor of power distribution lines. The necessary relevant parameters of the equivalent neutral conductor (R'_{eq} and r_{eq}) are obtained on the basis of currents measured in two feeding line phase conductors. These currents contain all relevant data, including data about the inductive influence of all, known and unknown, surrounding metal installations. Certainly, the actual reduction factor in the here-considered types of distribution lines involves, besides the inductive influence of the cable sheaths of cable section and ground wire of overhead section, also the influence of the grounding electrode at the place of their connection, that is, at the place of the discontinuity of the considered line. It is only necessary to mention that the test circuit used for this purpose should be performed by using the whole feeding line length, from the supply to the considered supplied substation. Only in this way, it is possible to take into account the influence of all relevant factors and parameters, including even those importance of which is negligible. Under this condition, it is possible to determine the actual ground-fault current distribution in any, including extremely complex situations, from the standpoint of surrounding metal installations.

4.4 Quantitative analysis

4.4.1 *Influence of cable section length*

This analysis is performed with the aim to assist the designers in developing an engineer's feeling concerning the influences of individual relevant factors and parameters influencing the value of the reduction factor of the here-considered lines. The necessary numerical data are adopted under the assumptions and approximations unavoidable at the design stage of a distribution substation and its feeding line.

At first, we shall assume that the substation is directly fed by a cable section with an uncoated metal sheath which is connected at the other end to an overhead line without any ground wire (Z_N and R_N are infinitely large). This situation is quite common in MV distribution networks with oil-filled, three-core cables. After the discretization of the cable sheath based on the following data: outer diameter $d = 44$ mm and experimentally determined $Z'_m/Z'_{sh} \approx 0.8$, and $Z_{sh} = (0.0007 + j0.002)$ Ω/m, the calculations performed based on the presented analytical expressions give the results that are graphically presented in Fig. 4.7.

As can be seen, when the cable section length L_c increases, the reduction factor r_i tends to the value that corresponds to the case when the cable is laid along the whole length of the feeding line.

Then, let us assume that a cable section directly supplies a 110/10-kV distribution substation, while, at the other end, the cable section is connected to the overhead section of the feeding line in the nearest transit substation. The cable has a lead sheath and its external and internal diameters are equal: 54.6 and 51.4 mm, whereas the value of the reduction factor of the cable line, according to the manufacturer's data, is equal to 0.22. For the overhead section of the line, we assume that in one case it is equipped with a steel ground wire so that it can be adopted that the reduction factor is: $r_0 = 1$, whereas in the other case we have a ground wire ACSR 95/55 mm^2 with a reduction factor: $r_0 = 0.675$. The values of the calculated reduction factor of the assumed lines versus cable section length for the first and the second case are shown in Figs. 4.8 and 4.9.

If we assume that the length of the cable section is 3 km (e.g., the average length of the cable line sections between two adjacent HV substations in the 110-kV network of Belgrade is approximately 4 km), the increase of reduction factor $|r_i|$ in comparison with the reduction factor of the line having only cable section (denoted

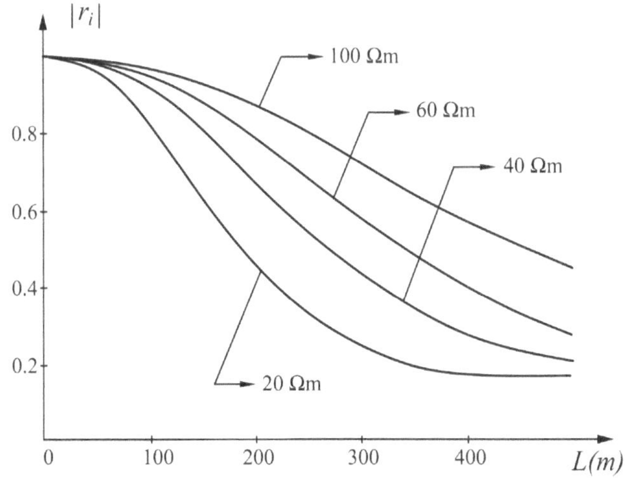

Figure 4.7 A reduction factor of a cable section with uncoated sheath and $r_0 = 1,\ Z_N = \infty,\ R_D = \infty$

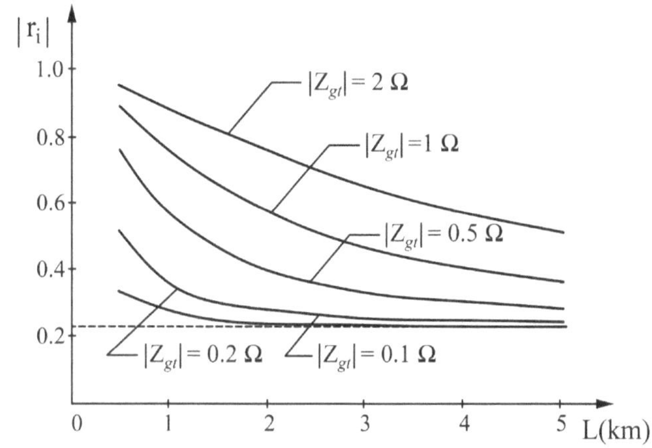

Figure 4.8 Cable with coated metal sheath and $r_0 = 1$.

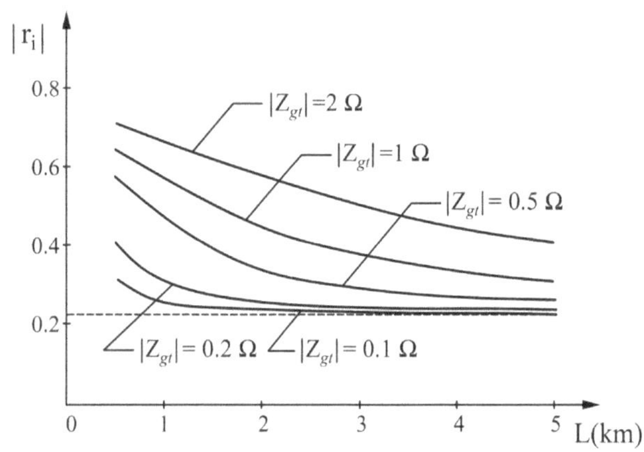

Figure 4.9 Cable with coated metal sheath and $r_0 = 0.675$.

with *dashed line*) is for the first case (Fig. 4.8) 6.4%, 13%, 49%, 110%, and 195%, whereas for the second case (Fig. 4.9), it is 4%, 10%, 32%, 71%, and 128% for $|Z_{gt}| = 0.1$, 0.2, 0.5, 1, and 2 Ω, respectively.

Based on the diagrams given in Figs. 4.8 and 4.9, it can be seen the following: when the value of impedance Z_{gt} is lower, the influence of the type of the ground wire, that is, the value of r_0, on the value of r_i is smaller. Also, when the cable section is longer, the influence of the type of ground wire of the overhead section, that is, the value of r_0, on the value of r_i, is smaller.

The effective value of the grounding impedance of substations in 110-kV distribution networks higher than 0.5 Ω ($|Z_{gt}| > 0.5$ Ω) is very rare in practice and may be expected only in the cases of very unfavorable (high) soil resistivity. However, according to the presented diagrams, one can also gain an insight into the value of the reduction factor when impedance Z_{gt} is extremely high, 1 and 2 Ω. In such cases the influence of the cable section length on the feeding line reduction factor is greater. Also, it is important to note that the greater part of the ground-fault current is dissipated into the earth through the grounding system of the transit substation. This effect is named a ground-fault current transfer and separately has been considered in Refs. [6,12].

On the basis of the presented calculation results, one can come to some conclusions important for the practical solving of the considered problem. By treating a feeding line composed of one cable and one overhead section as a cable line on the whole length results in an incorrect estimation of the achieved safety conditions in the supplied substation, so that these conditions are considered more favorable than they really are. This means that the prescribed safety criteria in some cases could be treated as satisfied, although this is not the case. This should be stressed because the standard testing measurements of safety conditions, which are mandatory before a newly built substation is put in operation, usually do not require verifications of the reduction factor value of the feeding line(s). This is the problem in the current practice that is separately considered in Chapter 6, Preliminary testing of safety conditions of high-voltage substations located in urban areas. Also, on the other side, treating such lines only as an overhead line may cause unnecessary costs in meeting the safety conditions in the supplied substation.

4.4.2 Lines consisting of different overhead sections

Let us consider the situation depicted in Fig. 4.2.(substation *A* is the only supply substation), where the new section of 110-kV line is equipped with two ground wires ACSR 95/55 mm^2, whereas the existing line has one ground wire, steel 50 mm^2. Also, we shall assume that the length of all line spans is 250 m and that the existing line section can be considered infinitely long from the standpoint of grounding effects of its ground wire. The longitudinal resistance of steel and ACSR ground wire is 3.5 and 0.299 Ω/km, respectively, whereas geometric mean radii are 3.49 and 6.23 mm, respectively. The line configurations for different sections are shown in Fig. 4.10.

Based on the presented analytical expressions, the performed calculations give results graphically presented in Fig. 4.11.

From the given diagrams, it is obvious that the considered effect is expressed more when it is more needed, that is, when the grounding impedance of the supplied substation, Z_b, is larger.

The considered cases of feeding lines are not too rare in contemporary power distribution practice and appear as a consequence of the significantly different degrees of urbanization of areas through distribution lines pass. The necessary calculations are relatively simple and take into account all relevant data available at the design stage. In other words, the presented analytical expressions have been developed on the basis of the approximations identical to those that we are otherwise forced to adopt at the design stage, because of the uncertainty and deficiency of some relevant data.

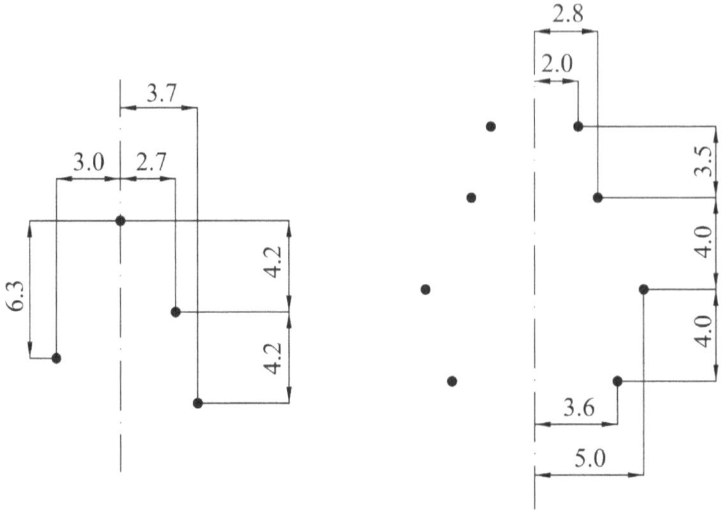

Figure 4.10 Disposition of line conductors of two different sections (distances in meters).

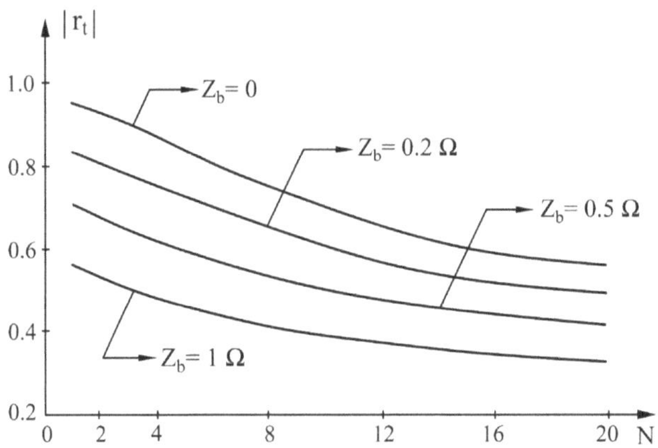

Figure 4.11 Influence of the number of spans with ACSR ground wires for different Z_b. *ACSR*, Aluminum conductor steel-reinforced wire.

As shown the presented analytical expressions enable at the design stage a necessary estimation of the ground-fault current distribution in substations supplied by the overhead feeding line consisting of sections with mutually different constructive characteristics. By developing the presented analytical expressions, the number of practically possible cases solvable by a relatively simple calculation procedure, that is, without the use of computers, has been enlarged in relation to the number of cases solvable by using the corresponding technical standard, for example [17].

References

[1] B. Thapar, S. Madan, Current for design of grounding systems, IEEE Transactions on Power Apparatus and Systems 100 (7) (1981) 342–345.
[2] Lj.M. Popović, Determination of the reduction factor for feeding cable lines consisting of three single-core cables, IEEE Transactions on Power Delivery 18 (3) (2003) 736–743.
[3] Lj.M. Popović, Improved analytical expressions for the determination of the reduction factor of the feeding line consisting of tree single-core cables, European Transactions on Electrical Power 18 (2) (2008) 190–203.
[4] Lj.M. Popović, Influence of the metallic installations surrounding the feeding cable line on the ground fault current distribution in supplied substations, IEEE Transactions on Power Delivery 23 (4) (2008) 2583–2590.
[5] Lj.M. Popović, Ground fault current distribution when a ground fault occurs in HV substations located in an urban areas, Progress in Electromagnetic Research B 59 (2014) 167–179.
[6] J.E.T. Villas, D. Mukhedkar, V.R. Fernandes, A.C. Magalhaes, Ground grid design of a transition station system—a typical example of fault transfer, IEEE Transactions on Power Delivery 5 (1) (1989) 124–129.
[7] S.T. Sobral, V.S. Costa, M.S. Campas, D. Mukhedkar, Dimensioning of nearby substations interconnected grounding system, IEEE Transactions on Power Delivery 3 (4) (1988) 1605–1614.
[8] Lj.M. Popović, Practical method for evaluating ground fault current distribution in station supplied by an unhomogeneous line, IEEE Transactions on Power Delivery 12 (2) (1997) 722–727.
[9] Lj.M. Popović, Efficient reduction of fault current through the grounding grid of substation supplied by cable line, IEEE Transactions on Power Delivery 15 (2) (2000) 556–561.
[10] Lj.M. Popović, Efficient reduction of fault current through the grounding grid of a substation supplied by an overhead line, European Transactions on Electrical Power 16 (3) (2006) 269–274.
[11] Lj.M. Popović, Ground fault current distribution in a station supplied by a line composed of a cable an overhead section, European Transactions on Electrical Power 17 (3) (2007) 207–218.
[12] S. Mangione, A simple method for evaluating ground-fault current transfer at the transient station of a combined overhead cable line, IEEE Transactions on Power Delivery 23 (3) (2008) 1413–1418.
[13] G. Zizzo, A. Campoccia, E. Riva Sanseverino, A new model for safety analysis in a system of earthing systems interconnected through bare-buried conductors, COMPEL Computation on Mathematics in Electrical Engineering 28 (2009) 412–436.
[14] Lj.M. Popović, Practical method for evaluating ground fault current distribution in station, towers, and ground wire, IEEE Transactions on Power Delivery 13 (1) (1998) 123–129.
[15] Lj.M. Popović, General equations of lumped parameter ladder circuit and special approach to analyzing electrical line transient states, International Journal of Electrical Power & Energy Systems 95 (2018) 568–576.
[16] Lj.M. Popović, Practical Methods for Analysis and Design of HV Installation Grounding Systems, Elsevier, Academic Press, 2018.

[17] Short-circuits currents in three-phase A.C. system – Part 3: Currents during two separate simultaneous line-to-earth short circuit and partial short-circuit flowing through earth. IEC Int. Std. 60909-3, (Ed.) 3.0, 2009-03.
[18] Lj.M. Popović, Active length of cable lines with uncoated metal sheaths as long grounding conductors, European Transactions on Electrical Power 22 (2) (2012) 206–215.

Measures for improving reduction factor of a feeding line

5

5.1 Overhead lines

5.1.1 Problem description

In unfavorable conditions and circumstances for solving the grounding problem of high-voltage (HV) distribution substations, the most acceptable solution can be obtained by undertaking special measures for reducing the ground-fault current fraction relevant for the ground grid potential rise. In the case of HV substations located in urban areas, the decisive reason for such measure is too high potential transferred to the nearest supplied medium-voltage (MV)/low-voltage substation, or too high potential in the incoming telecommunication circuits, railway lines, gas pipelines, etc.

In the general case, ground fault currents return to their sources in a power system through the earth as conductive media and neutral conductors of the feeding lines. The potential of a substation ground grid, as well as all potential differences relevant for safety conditions (step and touch voltages), appears as a consequence of grounding system impedance and the ground-fault current fraction that is dissipated from the substation grounding system into the surrounding earth, usually named earth current. Therefore for the designer of an HV substation, the possibility to accurately determine this current is of prime importance. One part of this current that is dissipated into the surrounding earth only through the ground grid of the supplied substation, the so-called grid current, is of special importance because of the following reason. The ground grid of an HV substation is usually the only grounding system element that is designed exclusively for the grounding purpose and as such the only one through which the designers can directly influence the grounding impedance of the whole grounding system [1]. When this possibility is already utilized, the possibilities for further reducing the grounding impedance of the whole grounding system formed in an urban area are very limited, or practically do not exist [1–9]. For certain, the lowest achieved value of this impedance, the maximum possible earth current, directly affects the grounding system characteristics expressed through step voltage, touch voltage, transferred potential, and ground potential rise. Therefore when we determined this current, we can also determine these characteristics of a grounding system and when they are within the prescribed framework, we say that the grounding system provides the necessary safety conditions, for example [10].

However, in the cases when conditions and circumstances for solving the grounding problem of a designed substation are unfavorable (too high soil

Relevant Characteristics of Power Lines Passing through Urban Areas. DOI: https://doi.org/10.1016/B978-0-323-91136-8.00001-3

resistivity, high ground-fault current level, a confined plots foreseen for the new substation, vicinity of public buildings, etc.), it may happen that the standard procedure of solving this problem is not sufficient. In such cases the solution can be obtained by undertaking special measures that lead to a general reduction of potentials and their differences on the whole grounding system, for example [11–13]. Generally speaking, this can be done by reducing the total ground-fault current or only its fraction dissipated through the grounding system of the supplied substation, that is, so-called earth current. However, since the reduction of total ground-fault current in practice is not feasible for only one (critical) substation, the only possibility is to reduce the earth current by changing ground-fault current relative distribution, that is, by improving the reduction factor of the feeding line(s).

Practical implementation of the measure for reducing earth current, in the case of HV substations, supplied through overhead lines is known under the name counterpoise and usually is described in the corresponding technical standards. However, some method for the calculation of the achieved quantitative effects of the application of this measure is not available in this literature.

An illustration of the counterpoise application is shown in Fig. 5.1, where source substation *A* supplies distribution substation *B* via an HV overhead line.

The used notation has the following meaning: *A* (*B*) is the supply (supplied) substation, $Z_a(Z_b)$ is the impedance of the grounding system of substation *A* (*B*), *D* is the place of discontinuity, d_{cc} is the diameter of the counterpoise conductor (buried bare copper wire), and *h* is the depth of the buried counterpoise conductor.

As can be seen, the counterpoise measure can be described as burying a bare copper wire of a certain length immediately under the overhead feeding line. The

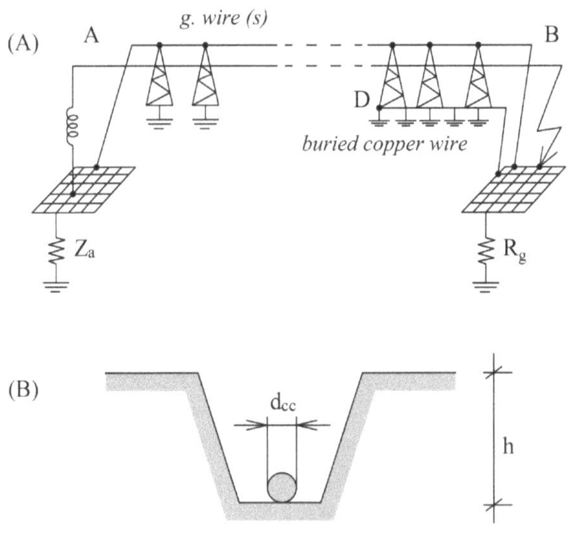

Figure 5.1 Practical application of the counterpoise (A) feeding line and (B) copper wire placed in the earth along an overhead line.

copper wire is conductively connected to the substation ground grid and the ground electrodes of a few close feeding line towers. In some critical cases from the standpoint of safety conditions, the reduction of the earth current can be the only acceptable solution for achieving required safety conditions. Besides, there are cases in practice, where the application of the counterpoint can be shown as the most economical solution, as well. The reasons for this can be already seen on the basis of Fig. 5.1 only. One of them is that the counterpoise conductor is further away from the elements of the ground grid than these elements are distant from each other. The second one is the fact that the current induced in the counterpoise conductor by the ground-fault current in the feeding line represents an additional diverted part of the ground fault current from the substation grounding grid and the whole grounding system.

According to Fig. 5.1, the measure of counterpoise is relatively easy for practical implementation, however, but the same cannot be said for the estimation of the effects achieved by this measure because they are a complex function of many different factors. For instance, the current dissipated through the substation ground grid into the surrounding earth produces a large potential field in the area around the substation. Since the counterpoise conductor is passing through this field, mutual interference (so-called proximity effect) between the ground grid and the counterpoise conductor in dissipating the ground fault current into the surrounding earth is unavoidable [7]. Also, the inductive coupling of this conductor with the feeding line conductors significantly contributes to the complexity of the electrical circuit that is formed in reality during a ground fault in the supplied substation.

Bearing in mind the previously mentioned facts, certain idealizations and simplification of the actual electrical circuit are necessary for developing a simple and for practical purposes sufficiently accurate analytical procedure for evaluating the effects of counterpoise application. The procedure that is directly (without using a computer) applicable and that at the design stage enables the estimation of the improved reduction factor of the feeding line achieved by using the counterpoise is developed in Ref. [11]. At this, it takes into account all more important factors and parameters under the assumptions and approximations usual at the design stage.

5.1.2 Calculated reduction factor

Let us assume that substation B is an outdoor HV distribution substation (Fig. 5.1). The difficulties in solving the grounding problem of such substations emanate from the immediate vicinity of public objects, as well as the fact that the overhead feeding lines, as is well known, have relatively high reduction factors. Also, the number of the external electrodes (outgoing MV cable lines, with uncovered or insulated metal sheath) can be relatively small, especially in the critical moment of putting a newly built substation into operation. As a result, the possibilities of attaining sufficiently low transferred potentials and required safety conditions on consumer premises can be very limited, especially in the case of too high local soil resistivity and a high level of ground-fault currents.

Based on Fig. 5.1, the counterpoise conductor represents an additional grounding conductor that provides one more return path for the ground-fault current. Namely, according to its spatial position, it has two different functions. One of them is a function of the grounding system external electrode, while the other one is the function of some type of shortened neutral conductor of the feeding line. The first one reduces the grounding system impedance, while the other reduces the current dissipated through the substation ground grid into the surrounding earth. Thus the only question is: How can one determine the current diverted from the ground grid of substation *B* by the counterpoise conductor?

Thus let us assume the conditions and circumstances where the grounding problem of the mentioned HV distribution substations appears in the most expressed form. Certainly, this is the case when the grounding system of an HV substation contains only the ground grid, that is, when all outgoing MV lines are foreseen as overhead lines without or with steel ground wire. These assumptions, at the same time, simplify the considered problem. However, according to the situation shown in Fig. 5.1, it is not difficult to imagine that despite these simplifications, a complex electrical circuit composed of many inductively and conductively connected elements will be formed during a ground fault in a supplied substation. With the aim of additionally simplifying the actual electrical circuit, we will make the following assumptions and approximations:

- The towers' footing electrodes along the counterpoise section do not exist.
- The counterpoise conductor is conductively connected to the overhead line ground wire only at its ends.
- Since in urban areas a correct determination of the equivalent soil resistivity is practically impossible, we assume that this soil is homogeneous and that its resistivity is somewhat higher than that approximately estimated for a maximum one.
- The currents in the neutral conductors along the counterpoise section (from *B* to *D*) do not influence the total ground-fault current, I_f.
- The current in the counterpoise conductor has a negligible influence on the current induced in the ground wire.

The last approximation is acceptable if the length of the counterpoise conductor L_{cc} is significantly smaller than the total line length L, or $L_{cc} < 0.2L$ that is a quite realistic assumption.

On the basis of the previous simplifications and by using

- the method of symmetrical components,
- the decoupled method, and
- the analytical expressions for reduction of an arbitrarily large ladder circuit to only one equivalent π-cell,

the grounding system shown in Fig. 5.1 can be presented by the relatively simple equivalent circuit shown in Fig. 5.2.

The used notation has the following meaning: I_f is the maximal anticipated ground-fault current component passing through the grounded neutrals in the supply network (known from the previous ground-fault current study); ΔI_i is the difference between the currents induced in the neutral conductors of the counterpoise section (ground wire and counterpoise conductor) and the current induced in the ground

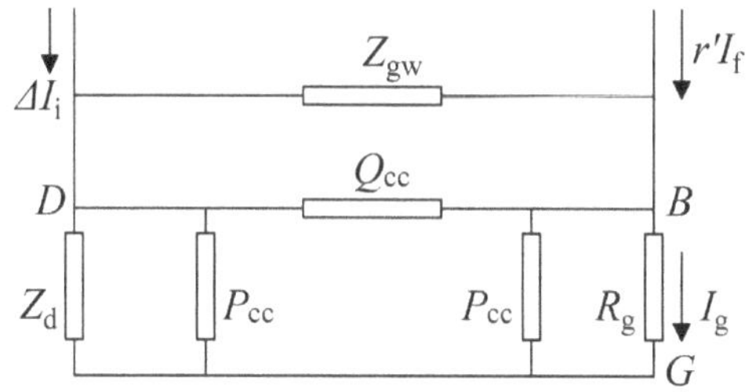

Figure 5.2 Equivalent circuit of the grounding system of substation *B*.

wire of the rest of the feeding line; I_g is the current dissipated into the ground through the substation ground grid; r' is the reduction factor of the feeding line, which involves the inductive influence of the counterpoise conductor under the assumption that it is laid along the whole line length; Z_{gw} is the self-impedance of the ground wire on the section with a counterpoise conductor; Z_d is the grounding impedance of the ground wire seen from discontinuity point *D* toward substation *A* (Fig. 5.1); Q_{cc} and P_{cc} are the impedances that make the equivalent π-cell of the lumped parameter ladder circuit formed by discretization of the counterpoise conductor; R_g is the resistance of the ground grid of the supplied substation; *D* is the discontinuity point, or end point of laying the counterpoise conductor; and *G* is the remote ground.

Then, let us notice the following: the total current appearing in the neutral conductors of the counterpoise section as a consequence of their inductive coupling with the faulty phase conductor is larger than the current appearing due to the same reason in the ground wire of the rest of the line. According to this, the current that represents the difference between these currents at the point of discontinuity, *D*, has to continue its flow toward the power system through the ground. This current denoted in Fig. 5.2 as ΔI_i can be, according to the definition of the reduction factor, for example [14], determined by

$$\Delta I_i = \left(\frac{Z'_m}{Z'_S} - \frac{Z'_{mC}}{Z'_{SC}}\right) I_f = (r_0 - r_c) I_f, \tag{5.1}$$

where Z'_S is the self-impedance of the feeding line ground wire(s) per unit length, Z'_m is the mutual impedance between the faulty phase conductor and the ground wire(s) per unit length, Z'_{SC} is the self-impedance of the neutral conductors of the counterpoise section per unit length, Z'_{mC} is the mutual impedance between the faulty phase conductor and the neutral conductors of the counterpoise section per unit length, r_c is the reduction factor of the line under the assumption that counterpoise is applied along the whole feeding line length, and r_0 is the reduction factor of the feeding line with no applied counterpoise.

Reduction factor r_c can be, according to the definition of this factor, determined under the assumption that R_g and Z_d are negligibly small ($R_g \approx 0$ and $Z_d \approx 0$). Then the equivalent circuit presented in Fig. 5.2 can be substituted by the simpler one shown in Fig. 5.3.

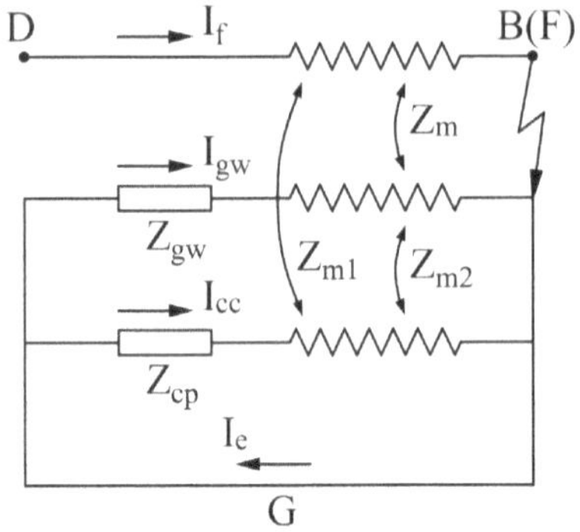

Figure 5.3 The currents induced in the neutral conductors on the counterpoise section.

The used notation has the following meaning: Z_{cc} is the self-impedance of the counterpoise conductor, Z_m is the mutual impedance between the ground wire and the faulty phase conductor, Z_{m1} is the mutual impedance between the counterpoise conductor and the faulty phase conductor, and Z_{m2} is the mutual impedance between the ground wire and the counterpoise conductor.

On the basis of the equivalent circuit in Fig. 5.3, one can write the following system of equations:

$$\begin{aligned} &Z_{gw}I_{gw} + Z_{m2}I_{cc} + Z_mI_f = 0, \\ &Z_{cc}I_{cc} + Z_{m2}I_{gw} + Z_{m1}I_f = 0, \\ &r_c = 1 + \frac{I_{gw} + I_{cc}}{I_f}. \end{aligned} \tag{5.2}$$

By solving this system of equations one obtains

$$r_c = 1 - \frac{Z_{m1}(Z_{gw} - Z_{m2}) + Z_m(Z_{cc} - Z_{m2})}{Z_{gw}Z_{cc} - 2Z_{m2}^2}. \tag{5.3}$$

Since we need the largest/critical value of current I_g, here it should be noted that the choice of the faulty phase conductor has to be done carefully. The phase conductor at the greatest mean geometrical distance from the ground wire and counterpoise conductors has not to be critical in all cases. Thus the calculation of the reduction factor should be performed for each of the line phase conductors with the aim to obtain its highest/critical value. Also, in many practical cases, in the rest of the line, the phase conductors are transposed and should be considered such in the determination of reduction factor r_0 [14].

Also, the necessary ground grid resistance in the majority of practical cases can be, according to Ref. [15], determined by

$$R_g = \rho_e\left[\frac{1}{l_g} + \frac{1}{\sqrt{20A}}\left(1 + \frac{1}{1 + h_g\sqrt{20/A}}\right)\right], \tag{5.4}$$

where R_g is the resistance of the substation grounding grid (Ω), ρ_e is the equivalent soil resistivity (Ωm), l_g is the total length of the buried conductor in meters (m), A is the area covered by the ground grid (m^2), and h_g is the depth of the ground grid in meters (m).

This expression can be used with reasonable accuracy for grid depths between 0.25 and 2.6 m.

According to Fig. 5.2, the counterpoise conductor is, from the standpoint of its grounding effects, presented by equivalent impedances Q_{cc} and P_{cc}. On the basis of the analogy of the counterpoise conductor (copper wire) and an uncovered cable sheath as grounding conductors, these impedances can be determined by using Eqs. (5.1) and (5.2), that is by

$$Q_{cc} = \frac{k^{L'_{cc}} - k^{-L'_{cc}}}{2} Z_{cc\infty}, \tag{5.5}$$

$$P_{cc} = \frac{k^{L'_{cc}} + 1}{k^{L'_{cc}} - 1} Z_{cc\infty}, \tag{5.6}$$

where L'_{cc} is the counterpoise conductor length expressed in relation to the length of 1 m, and $Z_{cc\infty}$ is the grounding (input) impedance of the counterpoise conductor seen from substation B and under the assumption that its length is infinite.

Variable k represents the current distribution coefficient in the nodes of the discrete parameter ladder circuit representing the counterpoise conductor as a grounding conductor under the assumption that its length is infinite ($L_{cc} \to \infty$). This coefficient is, according to Ref. [15], determined by

$$k = 1 + \frac{Z_{cc\infty}}{R_{cc}}, \tag{5.7}$$

where

$$Z_{cc\infty} = \frac{Z_{cc}}{2} + \sqrt{R_{cc}Z_{cc} + \frac{Z_{cc}^2}{4}}, \tag{5.8}$$

Variables Z_{cc} and R_{cc} in Eq. (5.8) denote discrete parameters of the counterpoise conductor for the discretization segment of 1 m.

If we now adopt that the length of the counterpoise conductor is approximately equal to its active length, from standpoint of grounding effects, instead of Eq. (5.8), one can use the following approximation:

$$Z_{cc\infty} \approx \sqrt{R_{cc}Z_{cc}}. \tag{5.9}$$

Then, under the assumption that the counterpoise conductor is approximately an equipotential electrode over its entire length, the grounding resistance of the copper wire of 1 m, according to Ref. [10], is

$$R_{cc} = \frac{\rho}{\pi m} \ln \frac{L_{cc}}{\sqrt{hd_{cc}}}, \quad \Omega, \tag{5.10}$$

where ρ is the equivalent soil resistivity around and along the counterpoise conductor (Ωm), L is the total counterpoise conductor length in meters (m), h is the depth in which the counterpoise conductor is laid in meters (m), d_{cc} is the outer diameter of the counterpoise conductor in meters (m), and m is the sign for the length of 1 m.

Also, self-impedance Z_{cc} of the counterpoise conductor for the segment of 1 m is

$$Z_{cc} = Z'_{cc}(\Omega/m) \times 1\ \text{m}, \tag{5.11}$$

where the self-impedance of the buried counterpoise conductor is, according to Ref. [14], determined by

$$Z'_{cc} = R'_{cc} + \omega\frac{\mu_0}{8} + j\omega\frac{\mu_0}{2\pi}\left(\frac{\mu_r}{4} + \ln\frac{\delta}{r_{cc}}\right)\ \Omega/\text{km}. \tag{5.12}$$

where R'_{cc} is the resistance of the counterpoise conductor (Ω/km); r_{cc} is the radius of the counterpoise conductor in meters (m); ω is the angular power frequency: $2\pi f$; μ_0 is the magnetic permeability of vacuum: $4\pi \times 0^{-7}$ Vs/Am; μ_r is the relative magnetic permeability of copper wire; and δ is the equivalent earth penetration depth in meters (m).

The impedance of the ground wire of the counterpoise section, Z_{gw} (Figs. 5.2 and 5.3), is obtained by

$$Z_{gw} = Z'_{gw}(\Omega/m) \times L_{cc}, \tag{5.13}$$

where Z'_{gw} is the longitudinal self-impedance of the ground wire per unit length. This impedance is also determined by Eq. (5.12) only instead of the parameters relevant for the counterpoint conductor we have the parameters relevant for the ground wire (R'_{gw} and r_{gw}).

Finally, grounding impedance Z_d is, according to Ref. [15], determined by

$$Z_d \approx Z_d(Z_A \approx 0) = \frac{k^N - k^{-N}}{k^N + k^{-N+1}} Z_\infty, \tag{5.14}$$

The coefficient of the current distribution along the infinitely long ground wire and the grounding impedance of such ground wire are determined by

$$k = 1 + \frac{Z_\infty}{R_t}, \tag{5.15}$$

$$Z_\infty = \frac{Z_S}{2} + \sqrt{R_t Z_S + \frac{Z_S^2}{4}}, \tag{5.16}$$

where N is the number of the spans of the feeding line from the point of discontinuity, D, to substation A; Z_S is the self-impedance of the ground wire(s) for one average span (Ω); and R_t is the average value of the tower footing resistance (Ω).

Now, when the analytical expressions for the determination of all elements of the circuit of Fig. 5.2 are known, the current I_g can be determined by the principle of superposition, that is, by the following relation:

$$I_g = I_g(r_c I_f) + I_g(\Delta I_i), \tag{5.17}$$

where $I_g(r_c I_f)$ is the component of current I_g created only by the ideal current source, $r_c I_f$ (Fig. 5.2); and $I_g(\Delta I_i)$ is the component of current I_g created only by the ideal current source, ΔI_i (Fig. 5.2).

In practical cases the lengths of HV overhead lines are more times larger than the active, from the standpoint of grounding effects, lengths of their ground wires. Thus in practical conditions, grounding impedance Z_d is independent of the length of the line section, from the point of discontinuity to the supply substation (N) and instead of Eq. (5.14) one can use the simpler expression (5.16).

The degree of reduction of current I_g obviously increases if the length of the counterpoise conductor increases. However, this reduction cannot be limitless. Any increase of the counterpoise conductor length over a certain limit value will not cause further reduction of current I_g. This limit length is the so-called active length of long grounding conductors and can be, according to Ref. [16], estimated by the following approximation:

$$L_a(\varepsilon = 0.05) \approx \frac{3m}{\ln\left(1 + \sqrt{\left(\pi |Z_{cc}| m / \rho \ln\left(L_a/\sqrt{hd_{cc}}\right)\right)}\right)}. \tag{5.18}$$

where L_a is the active length of the counterpoise conductor in meters (m), ε is the relative error, and m is the denotation for the length of 1 m.

To obtain the maximal possible effects of the counterpoise application, the length of the counterpoise conductor should be approximately equal to its active length. When this condition is satisfied, impedance P_{cc} becomes equal to $Z_{cc\infty}$ ($P_{cc} = Z_{cc}$), while impedance Q_{cc} becomes infinitely large ($Q_{cc} \to \infty$). Also, in this case, the current produced by the ideal current source ΔI_i through grounding resistance R_g becomes negligible. Thus the equivalent circuit of Fig. 5.2 becomes the simpler one as shown in Fig. 5.4.

The used notation has the following meaning: I_{gw} is the current generated by current source $r_c I_f$ in the ground wire(s), I_{cc} is the current generated by current source $r_c I_f$ in the counterpoise conductor, Z'_{gw} is the grounding impedance of the ground wire(s) seen from substation B, Z_{cp} is the grounding impedance of the counterpoise conductor involving proximity effect between it and the ground grid.

On the basis of the previously adopted length of the counterpoise conductor and the accordingly simplified electrical circuit, impedance Z'_{gw} is

$$Z'_{gw} = Z_{gw} + Z'_d, \tag{5.19}$$

where

$$Z'_d = \frac{Z_\infty Z_{cc\infty}}{Z_\infty + Z_{cc\infty}} \frac{I_{gw} + \Delta I_i}{I_{gw}}. \tag{5.20}$$

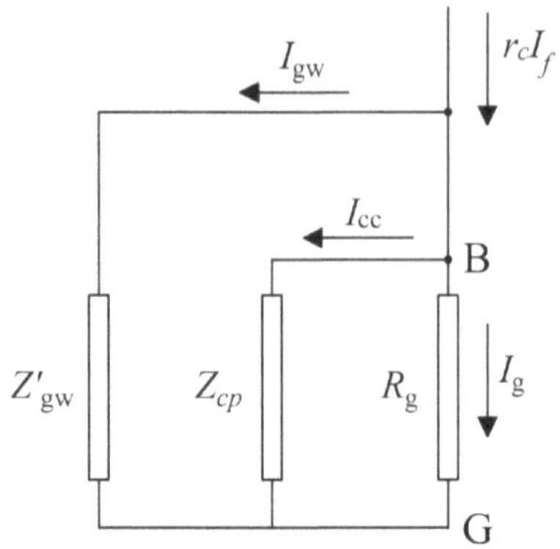

Figure 5.4 The equivalent circuit when $L_{cc} \approx L_{cc\infty}$.

On the basis of the circuit shown in Fig. 5.2, it is clear that current $r_c I_f$ is most intensively injected into the ground at the beginning of the counterpoise conductor, while current ΔI_i is most intensively injected into the earth at its end. According to this, the proximity effect between the ground grid and the counterpoint conductor influences only the distribution of current $r' I_f$ (Fig. 5.2). This effect can be taken into consideration by using, according to Ref. [7], the following analytical expression:

$$Z_{cp} \approx (1 + p) Z_{cc\infty} \tag{5.21}$$

where coefficient p expresses the influence of proximity between the grounding grid (electrode) and the counterpoise conductor. It is determined by

$$p = \frac{r_e Z_{c\infty}}{2R_S - (r_e - 1) Z_{c\infty}}. \tag{5.22}$$

Sign r_e represents the radius of the circle, surface of which is equal to the surface ground covered by the ground grid of the supplied substation determined by

$$r_e = \sqrt{\frac{A/m^2}{\pi}} \tag{5.23}$$

where A is the area covered by the ground grid (m^2).

In practical conditions the unfavorable effect of the proximity of the counterpoise conductor and the substation ground grid is certainly many times compensated by the favorable effect of metal installations that surround power lines in urban and suburban surroundings. Therefore before the final decision regarding the application of the considered measure, the actual reduction factor of the supply line should be determined. This is enabled by the methodological version developed in Refs. [17–19] and presented in Chapter 3, Actual reduction factor of power distribution lines.

5.2 Cable lines

5.2.1 Introduction

In urban areas, some of a large number of metal installations existing independently of our need for solving the grounding problem spontaneously take a function of grounding electrode of HV substations, while some others take also spontaneously the function of the neutral conductors of their feeding lines [1]. Thanking their influences, the grounding problem of HV distribution substations supplied through the cable lines becomes solvable in a huge number of practical cases without any grounding conductor laid in the earth only for the grounding purpose [1,15]. However, under extremely unfavorable conditions and circumstances, as they are: too high soil resistivity, high level of the ground-fault currents, a small number of metal installations surrounding the feeding line, a small number of outgoing cable lines (external grounding electrodes), cramped space foreseen for substation, and relatively old buildings (without foundation earth electrodes) in the substation surroundings, this problem can appear in a more pronounced form. This especially can be the case of HV distribution substations situated at the centers of consumption, that is, deep within the city area, where usually there are not sufficiently large plots available for laying necessary grounding electrodes and where there are many relatively old buildings (years of construction <1960) without foundation earth electrode. Under such conditions and circumstances, the substation grounding electrode potential rise can reach several kV and its reduction becomes a mandatory task for the designers of a certain substation. Thus the only solution to the grounding problem of such a substation can be undertaking a measure that reduces the fraction of the ground-fault current dissipated into the earth through the grounding system of this substation, the so-called earth current. Such a solution can be achieved by adopting a nonstandard HV cable construction with a sufficiently large cross section of metal sheath, that is, with a sufficiently low reduction factor of the feeding line. However, this solution may prove very expensive especially in the case where the cable line is relatively long.

The described problem can be efficiently overcome by laying a bare copper wire in the same trench with the cable line and by connecting it with the ground electrode of the supplied HV substation [12,13]. Certainly, such copper wire acts as an additional external electrode in relation to the substation grounding system. Besides that the additional fault current is diverted by this wire because of its intensive inductive coupling with the faulty phase conductor of the feeding line. Because of that the dissipation of the ground-fault current through this copper wire is as will be shown here more efficient than the enlargement of the cross section of the cable sheaths of the feeding line.

The practical application of this measure is illustrated symbolically by Fig. 5.5.

As well as in the previously considered case, in this case, it is necessary to determine the part of the ground-fault current appearing in the buried bare copper wire. One component of this current is induced by the ground fault current in the faulty phase conductor and at the point of line discontinuity (end of the buried copper

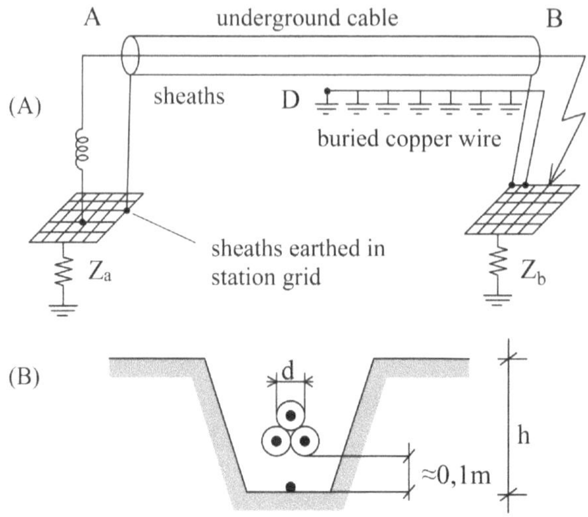

Figure 5.5 A cable line and buried copper wire: (A) cable feeding line and (B) cable and bare copper wire.

wire), *D* continues its flow toward its sources in the power system through the ground. The second component appears in the copper wire due to the conductive connection of this wire and the steel armature of the substation building foundation usually used as the main ground electrode of substations located in urban areas, for example [10].

The first developed method for solving this problem has been based on the too rough approximation that the current through the buried copper wire has no inductive influence on the currents through the cable line metal sheaths [12]. Because of that this method is improved in such a way that its present version by taking into consideration this fact enables higher accuracy in determining current I_b [13].

On the basis of Fig. 5.5, it is clear that the considered measure leads to a general reduction of potentials and potential differences appearing within and around an HV distribution substation by providing one additional return path for the ground-fault current. Thus the only question is the following: How to determine the quantitative effects of this special measure leading to improve the feeding line reduction factor?

By laying the copper wire on a certain length along the feeding cable line, this line cannot be anymore considered a line with uniformly distributed parameters along its length. Thus the existing analytical expressions, for the reduction factor of different cable lines, developed in Refs. [8,9] and presented in Chapter 2, Calculated reduction factor of feeding cable lines, cannot be applied.

Based on Fig. 5.5, it is also obvious that the buried copper wire has a dual function, that is, that it can be considered at the same time as an external electrode of the grounding system of substation *B*, as well as an additional and shortened neutral conductor of the feeding cable line. Thus the magnitude of this current for a certain value of current I_f depends on self- and mutual-impedances of the buried copper wire, faulty phase conductor, and cable sheaths of the feeding line, as well as the grounding impedance of the supplied substation, length of the buried bare cooper wire, local soil resistivity, etc.

5.2.2 Simplified equivalent circuit

For the beginning, let us assume that substation A is either source or transit one and that substation B is supplied by an HV cable line consisting of three single-core cables, as shown in Fig. 5.5. Also, let us assume that both of these substations belong to an HV network with directly grounded neutral points and that the current appearing in the foreseen copper wire has a negligibly small influence upon the value of the total ground-fault current, that is, that the length of the copper wire is relatively small in comparison with the feeding line length (Fig. 5.5).

With the aim to simplify the complex electrical circuit that is formed in reality during a ground fault in substation B, let us introduce an equivalent sheath substituting all three sheaths of a feeding line. The relevant parameters of the introduced equivalent sheath in the case of single-core cables laid in flat formation, according to Eq. (5.20), are

$$R'_{she} = \frac{R'_{sh}}{3} \tag{5.24}$$

$$r_{she} = \sqrt[3]{2r_{sh}d^2}, \tag{5.25}$$

whereas in the case of single-core cables laid in trefoil formation we have

$$r_{she} = \sqrt[3]{4}r_{sh}. \tag{5.26}$$

where R'_{sh} is the longitudinal cable sheath resistance (Ω/km), r_{sh} is the mean radius of cable sheath (m), and d is the distance between two adjacent single-core cables.

When relevant parameters of the introduced equivalent cable line sheath (R'_{she} and r_{she}) are known, by using the following theoretical bases:

- the method of symmetrical components,
- the decoupled method, and
- the reduction of uniform ladder circuits of an arbitrarily large number of π-cells to the one equivalent π-cell [15],

the grounding system shown in Fig. 5.5 can be presented by the relatively simple equivalent circuit shown in Fig. 5.6.

The used notation has the following meaning: U_1 is the voltage induced by current I_f in the equivalent cable line sheath; U_2 is the voltage induced by current I_i in the equivalent cable line sheath; U_3 is the voltage induced by current I_f in the buried copper wire; U_4 is the voltage induced by current I_{she} in the buried copper wire; I_f

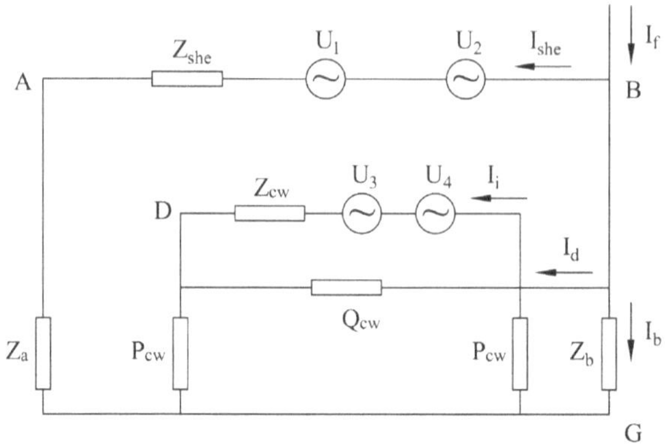

Figure 5.6 The equivalent circuit of the grounding system in Fig. 5.5.

is the anticipated maximal ground-fault current component returning into the power system through the grounded neutral points in the supply network; I_{she} is the ground-fault current through the equivalent cable line sheath; I_b is the ground-fault current dissipating into the earth through the substation grounding system excluding the buried copper wire; I_d is the ground-fault current diverted from the substation grounding system by the buried copper wire; I_i is the ground-fault current resulting from the influence of the inductive couplings between the buried copper wire on one side and the faulty phase conductor, including cable sheaths, on the other side; Z_{she} is the self-impedance of the equivalent cable line sheath; Z_{cw} is the self-impedance of the buried copper wire; Q_{cw} and P_{cw} are the impedances of the equivalent π-cell substituting the lumped parameter ladder circuit formed by discretizing of the buried copper wire as long grounding conductor; Z_a (Z_b) is the grounding system impedance of substation A (B); and G is the remote ground.

For the adopted relevant parameters of the buried copper wire, including its length and the equivalent resistivity of the surrounding soil, impedances: Z_{cw} is, according to Ref. [14], determined by

$$Z_{cw} = Z'_{cw}L = \left[R'_{cw} + \omega\frac{\mu_0}{8} + j\omega\frac{\mu_0}{2\pi}\left(\frac{\mu_r}{4} + \ln\frac{\delta}{r_{cw}}\right)\right]L, \quad \Omega, \tag{5.27}$$

where R'_{cw} is the copper wire resistance per unit length (Ω/km); r_{cw} is the radius of the copper wire (m); ω is the angular power frequency, $\omega = 2\pi f$; μ_0 is the magnetic permeability of vacuum: $4\pi \times 10^{-7}$ Vs/Am; μ_r is the relative permeability of copper wire; δ is the equivalent earth penetration depth (m); and L is the buried copper wire length (km).

On the basis of the adopted relevant parameters of the buried copper wire, the equivalent impedances Q_{cw} and P_{cw} are obtained by Eqs. (5.5)−(5.12).

Also, on the basis of previously determined parameters R'_{she} and r_{she}, for the certain cable line length, it is possible to determine impedance Z_{she} and all induced voltages: U_1, U_2, U_3, and U_4. Since impedances Z_a and Z_b can also be determined by using the corresponding procedure, for example [15], based on the equivalent circuit in Fig. 5.6, one can determine earth current I_b necessary for the estimation of the safety conditions within and in the vicinity of the supplied substation.

However, for determining the reduction factor in this special case of feeding cable line, it is necessary, according to the reduction factor definition (Chapter 2: Calculated reduction factor of feeding cable lines), to assume that impedances Z_a and Z_b are negligibly small ($Z_a \approx 0$ and $Z_b \approx 0$). Also, since we are interested in the largest possible effect of the considered measure, the length of the copper wire should be adopted in such a manner that is approximately equal to its active length from the standpoint of the grounding effects. According to this, impedance Q_{cw} can be considered infinitely large. After this approximation and the corresponding simplifications of the equivalent circuit in Fig. 5.6, we can write the following system of equations:

$$\begin{aligned} &Z_{she} I_{she} + Z_{m1} I_f + Z_{m2} I_i = 0 \\ &(Z_{cw} + P_{cw}) I_i + Z_{m3} I_f + Z_{m4} I_{she} = 0 \end{aligned} \tag{5.28}$$

where Z_{m1} is the mutual impedance between the faulty phase conductor and the equivalent cable line sheath, Z_{m2} is the mutual impedance between the equivalent cable line sheath and the buried copper wire, Z_{m3} is the mutual impedance between the faulty phase conductor and the buried copper wire, and Z_{m4} is the mutual impedance between the equivalent cable line sheath and the buried copper wire.

Bearing in mind the imagined physical appearance and position of the introduced equivalent sheath (Fig. 5.6), one can see that the impedances Z_{m2}, Z_{m3}, and Z_{m4} are mutually equal ($Z_{m2} = Z_{m3} = Z_{m4}$) and according to this for all of them can be used one the same symbol, Z_m. On the basis of the changed notations and Eq. (5.28), we obtain

$$I_{she} = \frac{Z_m^2 - (Z_{cw} + P_{cw}) Z_{m1}}{(Z_{cw} + P_{cw}) Z_{she} - Z_m^2} I_f, \tag{5.29}$$

$$I_i \approx \frac{Z_m Z_{m1} - Z_{she} Z_m}{(Z_{cw} + P_{cw}) Z_{she} - Z_m^2} I_f. \tag{5.30}$$

Then by using Eqs. (5.29) and (5.30), the improved reduction factor of the considered feeding line (Fig. 5.5) is obtained by

$$r_i = \frac{I_f - I_{she} - I_i}{I_f} \approx \frac{(Z_{cw} + P_{cw} - Z_m)(Z_{she} - Z_{m1})}{(Z_{cw} + P_{cw}) Z_{she} - Z_m^2} \tag{5.31}$$

Finally, on the basis of Eq. (5.31), the relative reduction of the feeding line reduction factor achieved by the buried copper wire, r_{cwi} is

$$r_{cwi} = \frac{r_i}{r_c} \approx \frac{(Z_{cw} + P_{cw} - Z_m)(Z_{she} - Z_{m1})}{\left[(Z_{cw} + P_{cw}) Z_{she} - Z_m^2\right] r_c}. \tag{5.32}$$

where r_c is the calculated reduction factor for the considered cable line without the buried copper wire determined by using analytical expressions corresponding to its specific spatial formation (Chapter 2: Calculated reduction factor of feeding cable lines).

Then, based on the approximation that the impedance Q_{cw} is infinitely large, the relevant reduction of current I_b resulting only as a consequence of the conductive connections of the cable line sheaths and the copper wire on one side and the substation grounding electrode on the other side, according to Fig. 5.6, is

$$r_{ct} = \frac{P_{cw}Z_{she}}{P_{cw}Z_{she} + Z_b(P_{cw} + Z_{she})}, \tag{5.33}$$

whereas the relative reduction only as a consequence of the conductive coupling between the buried copper wire and the substation grounding electrode is

$$r_{cwc} = \frac{P_{cw}}{P_{cw} + Z_b}. \tag{5.34}$$

Finally, the current dissipating through the grounding system of supplied substation B, excluding the buried copper wire, is determined by

$$I_b = r_t I_f = r_i r_{ct} I_f, \tag{5.35}$$

where r_t represents the coefficient of the total relative reduction of current I_b.

Also, the total effect of the copper wire on the relative reduction of current I_b is determined by

$$r_{cwt} = r_{cwi}\, r_{cwc}, \tag{5.36}$$

where r_{cwi} and r_{cwc} are determined by Eqs. (5.32) and (5.34).

In the case where the copper wire is laid along the entire cable line length, the given expressions should be modified. Then, in Eqs. (5.31) and (5.32) instead, P_{cw} should be Z_a, if substation A is transit one or should be omitted ($P_{cw} = 0$) if substation A is source one. Also, since the buried copper wire is long nearly as its active length (Eq. 5.18), impedance P_{cw} in Eqs. (5.33) and (5.34) can be substituted by impedance $Z_{cw\infty}$. Since dimensions of the grounding electrode of HV substations located in urban areas are relatively small, the proximity effect between the substation grounding electrode and buried copper wire can be disregarded [7,15].

Since cable lines are used almost without any exception in the urban and suburban areas, the value of equivalent soil resistivity of the surrounding area-earth cannot be exactly determined. Although there are several methods of measuring soil resistivity, none of them can be accurate in urban conditions. The objective obstruction represents the fact that surfaces of urban areas are covered by buildings, streets, pavements, and many other permanently constructed objects. Also, many known and unknown metal installations have already existed under the ground surface and affect measurement results. Because of that the equivalent soil resistivity can be only approximately estimated on the basis of the main geological characteristics of the soil through which the cable line will be laid. Here, it is important also to know the following: so far investigations of soil effects on the grounding impedance of long grounding conductors showed that the adoption of an equivalent soil resistivity somewhat higher (up to 20%) than the one approximately estimated ensures as actual one ensures the final results slightly on the side of the larger safety [15].

Because of all approximations and neglects done by developing the presented analytical expressions, complete insight and final judgment concerning the achieved safety conditions can be obtained only on the basis of test measurements that are, as it is well known, mandatory before putting a newly built substation into operation, for example [10].

5.3 Actual reduction factor

The previously considered special measures for improving feeding line reduction factors are used in cases of HV substations that are estimated as critical or unsatisfactory from the standpoint of safety criteria. However, these conditions can be accurately estimated only after taking into consideration favorable inductive influences of all, known and unknown, metal installations surrounding its feeding line passing through urban and/or suburban areas. Because of that, we can be quite sure of the necessity of the here-considered measures only if we have previously determined the actual reduction factor of the feeding line of the considered substation. The actual reduction factor of any feeding line passing through urban and/or suburban areas can be determined by using the methodology developed in Refs. [17–19] and presented in Chapters 3, Actual reduction factor of power distribution lines. However, for applying this methodology, it is necessary that we already have a constructed substation and its feeding line. As the time from the construction of the substation to its commissioning is relatively short, the time that can be used for the application of the here-considered measures is also short. However, the favorable circumstance is that the application of the here-considered measures does not require much time and because of that, they can be applied without postponing the term foreseen for putting a newly built substation into operation. In cases where they have been applied, it is necessary to once again determine the actual reduction factor of the feeding line, but this time by taking into account the effects of the applied measure. Only in this way, the effects of the application of the special here-considered measures for the improvement of the reduction factor of the feeding lines can be precisely determined.

The previously described procedure eliminates the possibility of applying the here-considered measures where they are not indispensable. It is quite realistic to assume that thanks to the influence of the surrounding metal installations, the actual picture about ground-fault current distribution will be so favorable that the measures to improve the reduction factors of feeding lines will be in many critical cases assessed as redundant.

When the here-presented measure is applied, the parameters of the equivalent neutral conductor (R'_{eq} and r_{eq}) determined by the methodology presented in Chapter 3, Actual reduction factor of power distribution lines, cumulatively involves the inductive influences of all, known and unknown, surrounding metal installations, including the bare copper wire laid with the aim to improve the reduction factor of the feeding cable line. It is only necessary to point out that the test circuit used for this purpose should be formed by using the whole feeding line length, from the supply to the considered supplied substation. Only in this way it is possible to take correctly into account the influence of all surrounding metal installations, including even those, influence of which is negligibly small. Under this condition, this methodology enables the correct determination of ground-fault current distribution for any HV or extra-high voltage (EHV) line in any, including extremely complex, situation from the standpoint of surrounding metal installations, that can be encountered in the central areas of the large urban agglomerations.

In practice, because of the inductive influence of surrounding metal installations, the favorable effect of the buried copper wire will be somewhat smaller, that is, it will be less pronounced than the one that is obtained by the here-presented analytical expressions. However, the complete situation concerning the safety conditions will be significantly better because the actual reduction factor of the considered cable line will be significantly lower (more favorable) than the calculated one. The favorable inductive influence of surrounding metal installations is greater in the cases where lines pass through the areas with a higher degree of urbanization because of a greater number of surrounding metal installations acting as additional ground-fault current return paths.

5.4 Quantitative analysis and general observations

5.4.1 Overhead lines

This analysis is performed with the aim to provide an insight into the quantitative side of the considered effects of the here-described measures for the reduction of the ground-fault current fraction dissipated through the ground electrode and grounding system of supplied HV distribution substation. Thus let us assume that substation *B* in Fig. 5.1 is supplied through an overhead 110-kV line equipped with an ACSR 95/55 mm^2 ground wire. Since this line is so long that it can be treated as infinitely long from the standpoint of grounding effects of its ground wire, impedance Z_d is equal to $Z_{gw\infty}$, or $(1.253 + j0.846)\ \Omega$. This value is used in all here-considered cases, independently of the adopted value of equivalent soil resistivity, ρ. Also, it is assumed that the line from point *D* toward substation *A* is with transposed phase conductors, whereas the height of the towers is 23.4 m, and the configuration of the conductors is shown in Fig. 5.7.

The counterpoise conductor is a copper wire of cross section 120 mm^2 laid in one case along two and in the other case along four nearest spans of an average length of 250 m and buried to a depth of 0.8 m. The ground grid is at the same

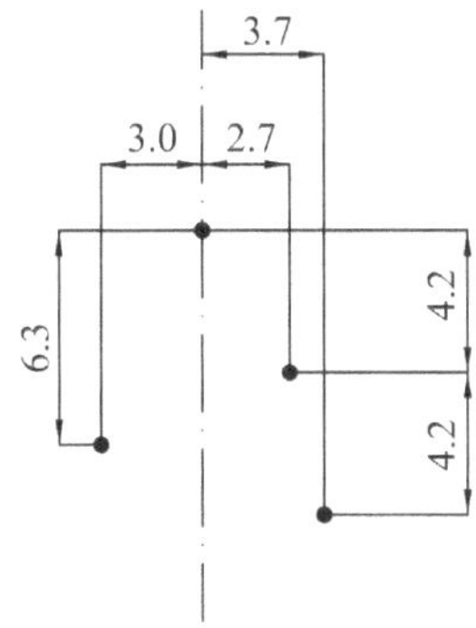

Figure 5.7 Disposition of the line conductors (distances in meters).

depth and has dimensions: 80 m × 100 m and spacing of 10 m. Calculations of the achieved grid current reduction have been performed for the following soil resistivity, 20, 50, 100, and 200 Ωm, covering the relatively wide scale of the practically possible values.

On the basis of the here-presented analytical expressions and the given data, it is obtained that the ground grid current is 31%, 25%, 20%, and 17% of current I_f for $\rho = 20$, 50, 100, and 200 Ωm, respectively. The fault current diverted from the ground grid and injected into the earth by the line ground wire and the counterpoise conductor is 37%, 41%, 44%, and 46% of current I_f, for $\rho = 20$, 50, 100, and 200 Ωm, respectively. The grid current without the application of the counterpoise would be 64%, 57%, 48%, and 37% of current I_f, for $\rho = 20$, 50, 100, and 200 Ωm, respectively.

If we bear in mind that the necessary length of the copper wire for the considered ground grid is 1780 m, according to the presented results, it is clear that the efficiency of the ground-fault current dissipation by the counterpoise conductor is significantly higher. Of course, this can be explained by the fact that this conductor acts as an external electrode in relation to the substation ground grid. Besides that the additional currents are absorbed by this conductor because of its inductive coupling with the faulty phase conductor of the feeding line and the conductive connections with the feeding line ground wire(s). Especially important is to note that the favorable effect of reduction of current I_g is more pronounced under less favorable natural conditions (higher ρ), that is, when the need for such effects is more pronounced.

The obtained results show that the measure of counterpoint should not be considered only as an exceptional measure that can be applied in cases where other possibilities for solving the problem are already completely utilized. In some cases can be shown that the most economical solution is obtained by restricting the length of the ground grid conductor to the required connections between the metal structures to be earthed and by applying the counterpoise. As shown here, it is more economical to invest in the counterpoise conductor than to further invest in the substation ground grid with the aim to reduce its grounding resistance.

5.4.2 Cable lines

Let us consider a 110-kV substation that is supplied through a cable line long 3 km. The foreseen cable type is XHE 94-A in the standard design, and its sheath has an outer diameter equal to 91 mm and is made of copper strings with a cross section equal to 95 mm^2.

The calculated reduction factor of this cable line when the equivalent soil resistivity is 30 Ω m and when its cables are laid in a trefoil formation is $r_c = 0.020 - j0.106$. Then, by selecting the next larger sheath cross section offered by the manufacturer's catalog equal to 120 mm^2, the value of the cable line–calculated reduction factor would be reduced to $0.015 - j0.084$. If a more reduction of current I_b is necessary, the next larger cross section is 185 mm^2 and the calculated reduction factor of the considered cable line would be $0.008 - 0j0.055$.

We will consider the effects of the here-described measure for improving the reduction factor of the cable line under the assumptions that the buried copper wire has a cross section of 120 mm^2, and that is buried in the depth of 1 m together with the line cables (Fig. 5.5). Also, we will assume that the equivalent soil resistivity is relatively high, or $\rho = 200$ Ωm, and that the effective value of grounding impedance Z_b is also relatively high, or $|Z_b| = 0.4$ Ω.

On the basis of the given data, by using Eqs. (5.32), (5.33), and (5.36), the variations of r_{cwi}, r_{cwc}, and r_{cwt} in a function of the copper wire length are graphically presented in Fig. 5.8.

According to the diagrams in Fig. 5.8, there is no purpose in laying the copper wire longer than the certain limiting length. As can be seen in the considered case, this length is about 1 km.

In the cases when the substation grounding system and substation main electrode (grid) have been already reinforced through the use of vertical earth rods, satellite ground grids [1], etc. so that a farther reduction of grounding impedance Z_b is not

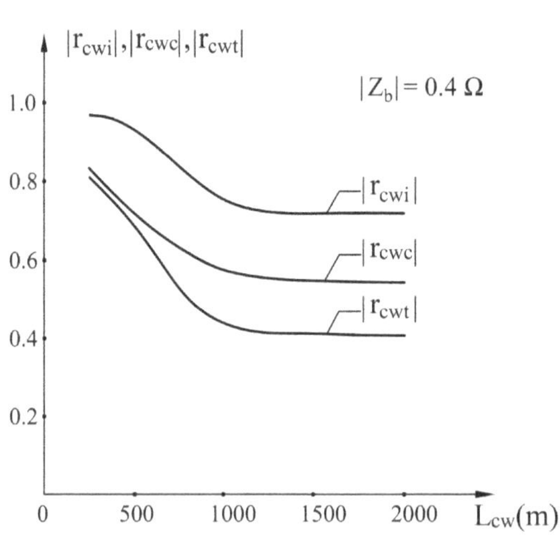

Figure 5.8 Coefficients r_{cwi}, r_{cwc}, and r_{cwt} depending on the copper wire length.

possible, the substation designers, in the general case, have only the following two possibilities:

- application of the measure considered herein, or
- placing an order for the manufacturers to fabricate a specially constructed cable having the sufficiently large cross-section of the metal sheath.

Since both of these technical possibilities provide the required effects, the following question becomes important. Which of these two technically feasible solutions is more cost-effective?

On the basis of the given data, it is not difficult to show that the reduction of 24.36% and 50.7% obtained by successive enlarging cable sheath cross section requires the additional consumptions of the copper strings that are 1990 and 7168 kg, respectively. However, for the reductions of 29.3% ($L = 500$ m), and 66.8% ($L = 1000$ m) obtained by the buried copper wire, it is sufficient only, 531 and 1062 kg, respectively. In practice, because of the inductive influence of surrounding metal installations, the favorable effects of the buried copper wire will be somewhat less pronounced. However, because of the inductive influence of surrounding metal installations, the favorable effects of the increased cross section of the cable sheaths will be also somewhat less pronounced. Thus the advantage of the here-considered measure is convincing.

The great efficiency of the measure considered here can be explained by the following facts. As a first, the strong inductive coupling between the buried copper wire and faulty phase conductors of the feeding cable line significantly intensifies the fault current dissipation into the ground through the buried copper wire. Second, the most intensive fault current dissipation occurs at the copper wire end distant from the substation grounding electrode so that the unfavorable so-called proximity effect practically does not exist.

On the basis of the presented quantitative analysis, it is much more economical to invest in the copper wire than to increase the cross section of the cable line metal sheaths. Also, this solution opens the possibility of applying the cheaper HV or EHV cables of standard construction in a whole distribution network, even in the cases where someone of supplied HV substations has extremely unfavorable conditions and circumstances for solving the grounding problem. Finally, it is important to mention that the effect of earth current reduction is more pronounced when it is more needed, that is, when the values of soil resistivity and grounding impedance Z_B are higher.

The developed analytical expressions enable, at the design stage, a relatively easy and reasonably accurate estimation of the reduction degree of the ground-fault current fraction dissipated through the grounding system of a certain HV distribution substation. In practice, the achieved effects will be somewhat smaller due to the inductive influence of the surrounding metal installations, but the overall situation concerning the safety conditions within and in the vicinity of the supplied substation will be more favorable than that obtained by calculations. How much the actual safety conditions will be more favorable depends on the degree of urbanization of the area through which the feeding line passes and determined by applying the methodology presented in Chapter 3, Actual reduction factor of power distribution lines.

References

[1] Lj.M. Popović, Comparative analysis of grounding systems formed by MV cable lines with either uninsulated or insulated metal sheath(s), Electric Power System Research 81 (2) (2011) 393–399.

[2] S.T. Sobral, et al., Ground potential rised of urban step-down substations fed by power cables—a practical example, IEEE Transactions on Power Delivery 3 (4) (1988) 1564–1572.

[3] J.E.T. Villas, D. Mukhedkar, V.R. Fernandesm, A.C. Magalhaes, Ground grid design of a transition station system—a typical example of fault transfer, IEEE Transactions on Power Delivery 5 (1) (1989) 124–129.

[4] G. Zizzo, A. Campoccia, E. Riva Sanseverino, A new model for safety analysis in a system of earthing systems interconnected through bare-buried conductors, COMPEL – The International Journal for Computation and Mathematics in Electrical Engineering 28 (2009) 412–436.

[5] E. Pons, P. Collela, R. Tommasini, R. Napoli, P. Montegiglio, G. Gafaro, et al., Global earthing system: can buried metallic structures significantly modify the ground potential profile? IEEE Transactions on Industry Applications 51 (6) (2015) 5044–5049.

[6] A. Meliopulos, R. Web, E. Joy, S. Patel, Computation of maximum earth current in substation switchyards, IEEE Transactions on Power and Systems PAS-102 (9) (1983) 3131–3139.

[7] Lj.M. Popović, Proximity effect between an earthing grid and external electrodes in an earthing system of high-voltage installations, IEE Proceedings C 133 (6) (1985) 346–352.

[8] Lj.M. Popović, Determination of the reduction factor for feeding cable lines consisting of three single-core cables, IEEE Transactions on Power Delivery 18 (3) (2003) 736–743.

[9] Lj.M. Popović, Improved analytical expressions for the determination of the reduction factor of the feeding line consisting of tree single-core cables, ETEP: European Transactions on Electrical Power 18 (2) (2008) 190–203.

[10] IEEE Guide for Safety in Substations Grounding, Std. 80-2000.

[11] Lj.M. Popović, Efficient reduction of fault current through the grounding grid of a substation supplied by an overhead line, ETEP: European Transactions on Electrical Power 16 (3) (2006) 269–274.

[12] Lj.M. Popović, Efficient reduction of fault current through the grounding grid of substation supplied by cable line, IEEE Transactions on Power Delivery 15 (2) (2000) 556–561.

[13] Lj.M. Popović, Reduction of the fault current passing through the grounding system of an HV substation supplied by cable line, International Journal of Electrical Power & Energy Systems 99 (2018) 493–499.

[14] Short-circuits currents in three-phase A.C. system – Part 3: Currents during two separate simultaneous line-to-earth short circuit and partial short-circuit flowing through earth, IEC Int. Std. 60909-3, (Ed.) 3.0, 2009-03.

[15] Lj.M. Popović, Practical Methods for Analysis and Design of HV Installation Grounding Systems, Elsevier Academic Press, 2018.

[16] Lj.M. Popović, Active length of cable lines with uncoated metal sheaths as long grounding conductors, ETEP: European Transaction on Electrical Power 22 (2) (2012) 206–215.

[17] Lj.M. Popović, Ground Fault current distribution when a ground fault occurs in HV substations located in an urban areas, Progress in Electromagnetic Research B 59 (2014) 167–179.
[18] Lj.M. Popović, Determination of the actual reduction factor of HV and MV cable line passing through urban and suburban area, Facta Universitatis, Series: Electronics and Energetics 27 (1) (2014) 25–39.
[19] Lj.M. Popović, Preliminary testing and assessment of safety conditions of HV substations located in urban areas, Electric Power System Research 195 (2021) 107111. Available from: https://doi.org/10.1016/j.epsr.2021.107111.

Preliminary testing of safety conditions of high voltage substations located in urban areas

6

6.1 Introduction

As is well known the final judgment concerning the safety conditions within and in the vicinity of a certain high voltage (HV) substation can be obtained only by test measurements. These measurements are performed immediately before putting into operation the newly built HV substation and periodically with the aim of verifying the required safety conditions. Only after that we can be sure that the newly built substation will be fit for its intended purposes under all practically possible operating conditions. Because of that these test measurements are required by the relevant technical standards and apply to all HV substations regardless of where they are located, for example, [1]. However, in a relatively large number of HV substations located in the urban and suburban areas, they cannot give the real picture concerning the achieved safety conditions because we did not know until recently the actual reduction factor of HV lines that pass through urban and/or suburban areas [2,3]. As is known the reduction factor represents the ratio of the fraction of the ground-fault current that is dissipated into the surrounding earth through the grounding system of the supplied substation and the total ground-fault current, for example, [2].

According to contemporary practice in electric power distribution companies, the grounding system characteristics of an HV/medium voltage (MV) substation are tested by using a version of the "fall-of-potential" method adapted for conditions and circumstances in HV distribution networks. The basic idea used for this purpose is injecting a test current through the grounding system of the tested substation via one phase conductor of the feeding line and by using the grounding system of the source substation as an auxiliary grounding electrode. If the source substation is not the adjacent one, then, via the grounding system of the nearest transit substation supplied by the same feeding line. In these adjacent (source or transit) substations, the transformers that serve their own consumption are usually used as a voltage source for the test circuit formed in this manner. The necessary potential measuring circuit is formed by using a provisory potential electrode positioned on the "remote earth" and as much as possible by the rules defined for this circuit, for example, [4,5]. The "remote earth" can usually be found at an approximately radial distance, which is between 300 and 500 m from the tested HV/MV substation. However, in areas with high values of soil resistivity, this distance can be significantly greater [6,7].

Relevant Characteristics of Power Lines Passing through Urban Areas. DOI: https://doi.org/10.1016/B978-0-323-91136-8.00006-2

Using the previously described procedure, it is possible to measure the potential and maximal potential gradients created by the test current on the grounding system of the tested HV/MV substation. However, by this procedure, the real picture concerning the safety conditions inside and in the vicinity of a tested HV/MV substation can be obtained only if the test current circulates over the entire length of the feeding line, that is, from the source to the tested substation [3,8]. The problem arises because this requirement is not easy to fulfill in many practical cases. These are the cases where one or more transit substations exist between the tested and source substation. These substations should be temporarily out of operation but it is very difficult to obtain permission to interrupt the power supply to many consumers, including those that are essential for the normal functioning of public services in a big contemporary city. Therefore as a rule, the nearest transit substation is used as an auxiliary grounding electrode in forming a necessary test circuit.

The latest attempts to solve the described problem are presented in the papers of Refs. [9,10]. However, the developed methods based on the shortened test current circuit give too conservative estimations. As such, these methods are not desirable in cases where the achieved safety conditions are such that the relevant potentials and potential differences are close to their maximal permissible value. Also, the problem was considered and solved in principle by developing the test procedure for the preliminary estimation of prescribed safety conditions [3]. Before that it was necessary to develop the method that enables the determination of actual ground fault current distribution in the case of a fault within an HV substation located in the urban environment. The procedure for the determination of the actual distribution of ground-fault current, in that conditions, was developed on the basis of the experimental investigations performed in the HV distribution network of Beograd and the analytical procedure developed by introducing a fictitious (equivalent) conductor [3,8]. This conductor was imagined as a cylinder along and around the feeding line conductors and its relevant parameters are the longitudinal resistance and mean radius. They are determined under the condition that this conductor substitutes all, known and unknown, metal installations existing around and along the feeding line, from the standpoint of their inductive influence on ground-fault current distribution. However, this method is not applicable in the cases of feeding lines with applied cross-bonding that is the most common case in the contemporary electric power distribution practice [11].

The here-presented methodology version is modified in relation to the version presented in Ref. [3] in such a way that can be applied for any type of HV or extra high voltage (EHV) feeding lines. The input data for obtaining the actual grounding system impedance are the measured simulated ground-fault current through one of the line phase conductors and the measured fraction of this current induced in one phase conductor additionally used for obtaining the here-defined test circuit. When the value of the actual grounding system impedance is determined, it is possible to conduct the correct preliminary estimation of safety conditions within and in the vicinity of the tested HV substation. Since the grounding system impedance of HV substations located in urban areas is usually very low [6,12], it is quite realistic to assume that the preliminary estimation made by the test procedure presented here will be sufficient for a reliable definite judgment about safety conditions in a large number of HV substations located in urban and suburban area.

6.2 Problem description

As is shown in the previous chapters for safety conditions within and in the vicinity of an HV substation is relevant only the fraction of the total ground-fault current dissipated through the substation grounding system into the surrounding earth or the so-called earth current. Thus for the correct estimation of safety conditions, it is of prime importance to correctly determine this fraction of the ground-fault current. However, this is a complex problem in the case of substations located in urban and suburban areas. Grounding systems of HV distribution substations consist of a grounding electrode and many outgoing MV cable lines acting as external grounding electrodes and/or conductive connections with the grounding systems of the supplied MV/low voltage (LV) substations [6,7]. These grounding systems include, through the terra-neutral (TN) earthing system in the LV network and consumer installations, many known and unknown, metal installations typical for the urban areas [13,14]. As a result, outgoing cable lines simultaneously become the conductive connections with the metal installations laid along the same street(s) as the feeding line of an HV substation. Due to this, the currents appearing during a ground-fault in the metal sheaths of each outgoing cable line depend on many local factors, inter alia, on the structure and spatial disposition of metal installations surrounding the considered HV substation as well as on the structure and spatial disposition of metal installations surrounding its feeding line. Thus it is not difficult to see that the current passing through the grounding system elements only as a consequence of the inductive influence of surrounding metal installations is distributed on many different flows in each HV substation.

Nevertheless, on the huge number of different small flows, the fault current passing through the elements of the grounding system of an HV substation has only two, essentially different fractions. One of them is injected into the surrounding earth, while the other is induced in the metal installations surrounding the feeding line. Since the split between these two fractions occurs along the many grounding system elements located under the surface of the ground, none of these two fractions can be singled out and separately determined by direct or indirect measurements. However, without the known value of the current injected into the surrounding earth, the determination of the actual grounding system impedance of the tested HV substation is not possible.

According to the previous consideration, in practical conditions, besides the used conductors of the feeding line, many other metal installations typical for urban areas spontaneously participate in forming the test circuit for an HV substation. These installations are mainly situated along the same street(s) as a feeding line section used for obtaining the test circuit. Some of them are in direct and continuous contact with the earth, whereas some others are insulated toward the earth.

Also, many of these installations are grounded and mutually connected by applying the TN earthing system in almost every building along the streets of contemporary cities. Thus independent of their basic functions these installations spontaneously form one large and unique network of different mutually connected electrical conductors. The structure and topology of this mainly underground network are usually very complex, depend on many local conditions and circumstances, and cannot be visually determined or verified.

The problem was solved by defining the corresponding test measurements and by introducing an equivalent conductor substituting, from the standpoint of their inductive influence, all surrounding metal installations [8]. At this, it was utilized the favorable fact that currents induced in the surrounding metal installations circulate only in the axial direction through them and as such cannot create any potential toward the ground [3,8,15–18]. Due to this, it is not relevant whether these installations toward the ground are insulated or not, and whether the introduced equivalent conductor substituting all these installations has any insulation, or not. We are also using this fact when we determine the reduction factor of HV overhead lines. Namely, in that case, we also do not take into account the fact that their ground wires are grounded at each of their towers, for example, [2]. Also, for developing the whole Popović's methodology and its application, the relevant data about the constructive characteristics and mutual spatial position of surrounding metal installations as well as their spatial positions toward the considered HV or EHV feeding line are not necessary. Instead of these data, for the application of this methodology are necessary and sufficient only two currents that are determined by test measurements. They enable taking into consideration all, known and unknown, relevant data in all practically possible situations from the standpoint of the urban environment of the line. In the methodology version in Ref. [3], these two currents are the measured values of the test current in the phase conductor of one of the line single-core cables and its fraction induced in the metal sheath of the same that cable. However, the procedure developed in Ref. [3] refers only to the case when the feeding line is an HV cable line with no applied cross-bonding. Since the application of such cable lines is suggested only for very short lines, for example, [11,19], their application is not usual in contemporary HV distribution networks. Thus in further research work, it was necessary to enlarge the field of the methodology application so that it can be used for all cases of HV or EHV feeding lines passing through urban and/or suburban areas.

The mentioned problem has been overcome by introducing a certain improvement in the methodology version developed in Ref. [3]. Instead of one, in the general case, neutral conductor, one more phase conductor (in the further consideration named additional phase conductor) is used to form the necessary test current circuit. In this way the problem has been solved so that this methodology version can be used for any type and any constructive characteristic of an HV or EHV line [18]. Also, due to the significantly higher conductivity of phase conductors in relation to neutral conductors of any power line, the methodological version presented here is more accurate in comparison with the previous version based on measuring the current induced in one of the neutral conductors of the considered line.

6.3 Relevant parameters of the introduced equivalent sheath

6.3.1 *Cable lines with applied cross-bonding*

In the beginning, let us assume that the surrounding metal installations do not exist. Then, the test circuit simplified by this assumption can be presented as shown in Fig. 6.1.

The used notation has the following meaning: *A* (*B*) is the transit (tested) substation; U_a is the auxiliary voltage source; I_t is the test (simulated ground-fault) current; I_1 is the current induced in the additional phase conductor; I_e is the current injected into the earth through the grounding system of substation *B*; *A* is the ampere-meter; and *D* is the disconnection that should be in the open position during the measurements.

Metal sheaths of the cable line are excluded from the test circuit because the tests are performed only to determine the actual impedance of the grounding system of the tested substation.

To simplify the problem, all known and unknown metal installations participating in test current underground distribution are substituted by only one equivalent conductor. This conductor is imagined as a cylinder around and along the feeding line section used for the test circuit and can be considered an additional neutral conductor (or metal sheath) of the cable line on this section. By introducing this conductor into the measurement circuit in Fig. 6.1, the actual test circuit can be substituted by a relatively simple equivalent circuit in Fig. 6.2.

The used notation has the following meaning: I_i is the current induced in the equivalent conductor/sheath; Z_{ph} is the self-impedance of the additional phase conductor; Z_{eq} is the self-impedance of the introduced equivalent conductor/sheath; Z_{01} is the mutual impedance between the phase conductor with current I_t and the additional phase conductor with current I_1; Z_{0eq} is the mutual impedance between the phase conductor with current I_t and the equivalent conductor/conductor; Z_{1eq} is the mutual impedance between the additional phase conductor and the equivalent conductor/sheath; and Z_A (Z_B) is the grounding system impedance in substation *A* (*B*).

The relevant parameters of the equivalent sheath can be determined by using the condition that the currents I_t, I_1, and I_e (Fig. 6.1) have to remain unchanged after introducing the equivalent sheath. On the basis of the equivalent circuit in Fig. 6.2

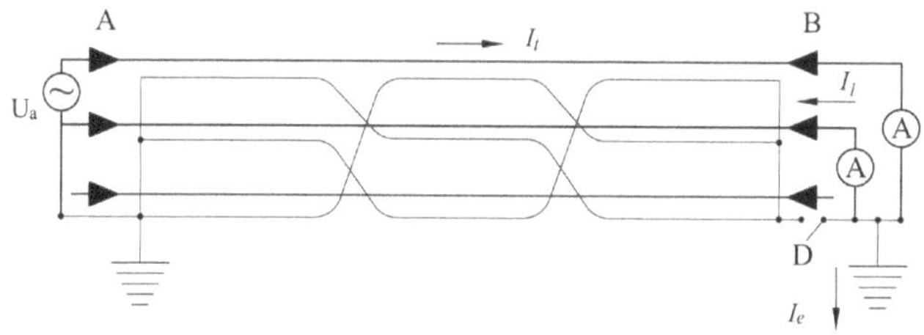

Figure 6.1 The principal measurement circuit.

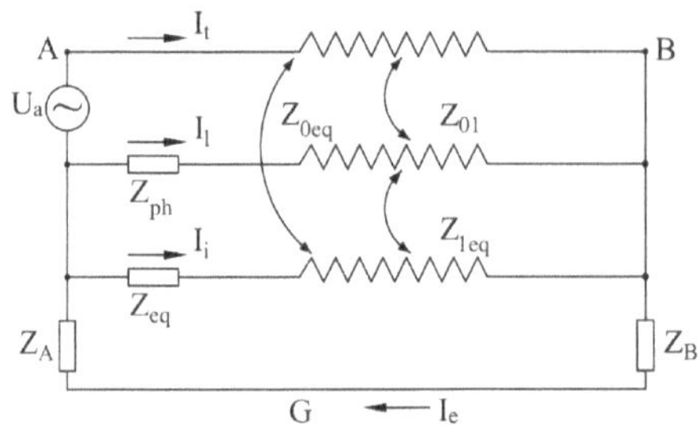

Figure 6.2 Simplified equivalent circuit of the actual test circuit.

and approximations $Z_A \approx Z_B \approx 0$, this condition can be expressed by the following system of equations:

$$\begin{aligned} &1.\ Z'_{01}I_t + Z'_{ph}I_1 + Z'_{1eq}I_i = 0 \\ &2.\ Z'_{0eq}I_t + Z'_{1eq}I_1 + Z'_{eq}I_i = 0 \end{aligned} \tag{6.1}$$

Impedances Z'_{ph} and Z'_{01} are, according to Ref. [2], determined by

$$Z'_{ph} = R'_{ph} + \omega_t \frac{\mu_0}{8} + j\omega_t \frac{\mu_0}{2\pi}\left(\frac{\mu_r}{4} + \ln\frac{\delta_t}{r_{ph}}\right), \quad \Omega/\text{km}, \tag{6.2}$$

$$Z'_{01} = \omega_t \frac{\mu_0}{8} + j\omega_t \frac{\mu_0}{2\pi}\ln\frac{\delta_t}{d}, \quad \Omega/\text{km}, \tag{6.3}$$

where R'_{ph} is the longitudinal resistance of the phase conductor, (Ω/km); r_{ph} is the radius of the phase conductor (m); ω_t is the angular test circuit frequency: $2\pi f_t$; μ_0 is the magnetic permeability of vacuum: $4\pi \cdot 10^{-7}$ Vs/Am; μ_r is the relative magnetic permeability of phase conductor; and d is the distance between two adjacent phase conductors (single-core cables).

The equivalent earth penetration depth, δ_t, is determined by

$$\delta_t = 658\sqrt{\frac{\rho}{f_t}}(\text{m}), \tag{6.4}$$

where ρ is the equivalent soil resistivity along and around the cable line (Ω m), and f_t is the test circuit frequency (somewhat higher than power frequency, 50 or 60 Hz).

According to the imagined physical appearance and spatial position of the equivalent sheath, the used section of the feeding line, including all surrounding metal installations can be during the test measurements presented by a relatively simple line model, cross section of which is shown in Fig. 6.3.

The additional phase conductor is marked differently (*lighter shade*) to be shown that for the need of test measurements, it has a function of one neutral conductor.

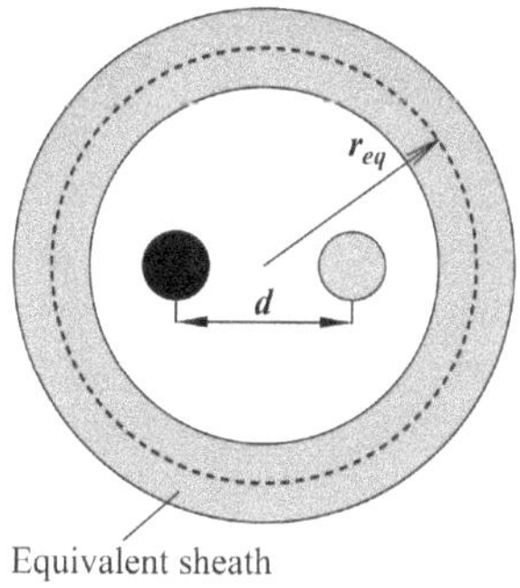

Figure 6.3 Cross section of the simplified line model.

In the here-considered case the equivalent sheath, according to Fig. 6.1, does not encompass the influence of the metal sheaths of the considered line.

Since the introduced equivalent conductor is imagined like a joint sheath for the two applied phase conductors, the analytical expressions for Z_{eq}, Z_{0eq}, and Z_{1eq} are a priori known and, according to Ref. [2], they are

$$Z'_{eq} = R'_{eq} + \omega_t \frac{\mu_0}{8} + j\omega_t \frac{\mu_0}{2\pi} \ln \frac{\delta_t}{r_{eq}}, \quad \Omega/\text{km}, \tag{6.5}$$

$$Z'_{0eq} = Z'_{1eq} = \omega_t \frac{\mu_0}{8} + j\omega_t \frac{\mu_0}{2\pi} \ln \frac{\delta_t}{r_{eq}}, \quad \Omega/\text{km}, \tag{6.6}$$

where R'_{eq} is the resistance of the equivalent sheath per unit length (Ω/km), and r_{eq} is the mean radius of the equivalent sheath (m).

Since currents I_t and I_1 are obtained by measurements, they can be considered priory known quantities, and condition defined by Eq. (6.1) can be modified into

$$\frac{Z'_{0eq}Z'_{1eq} - Z'_{eq}Z'_{01}}{Z'_{eq}Z'_{ph} - Z'^2_{1eq}} = \frac{I_1}{I_t} \tag{6.7}$$

Also, since Eq. (6.7) gives the relationship between complex quantities, it can be presented as the following system of equations:

$$\begin{aligned} \text{Re}\left\{\left(Z'_{0eq}Z'_{1eq} - Z'_{eq}Z'_{01}\right)I_t\right\} &= \text{Re}\left\{\left(Z'_{eq}Z'_{ph} - Z'^2_{1eq}\right)I_1\right\}, \\ \text{Im}\left\{\left(Z'_{0eq}Z'_{1eq} - Z'_{eq}Z'_{01}\right)I_t\right\} &= \text{Im}\left\{\left(Z'_{eq}Z'_{ph} - Z'^2_{1eq}\right)I_1\right\}, \end{aligned} \tag{6.8}$$

where Re and Im are the real and imaginary parts of this equation.

In Eq. (6.8) the only unknown variables, according to Eqs. (6.2), (6.3), (6.5), and (6.6), are R'_{eq} and r_{eq}. Thus by solving this system of equations, we can determine the relevant parameters of the equivalent sheath.

Because of the previous approximations, $Z_A \approx Z_B \approx 0$, it is necessary to mention that HV distribution substations are mainly located in urban or suburban areas where, as is well known, many underground metal installations act as perfect grounding electrodes [6,7]. Thus the quantitative relationships between the impedances shown in the equivalent circuit, Fig. 6.2, are such that grounding impedances Z_A and Z_B can be disregarded. On the basis of these approximations, each current appearing in the additional phase conductor and any of the surrounding metal installations is a consequence solely of the mutual inductive coupling with each of all other surrounding electrical circuits as shown in Chapter 3, Actual reduction factor of power distribution lines. However, the whole methodology is developed on the basis of the measured currents I_t and I_1. According to that it takes into account the fact that the real values of Z_A and Z_B are somewhat larger than zero. Because of that currents I_t and I_1 contain components appearing only as a consequence of potentials $Z_A I_e$ and $Z_B I_e$ at the ends of the test circuit. However, these components of currents I_t and I_1 are negligibly small in practical conditions because of the following reasons:

- current I_e is only a few percent of I_t because of the high conductivity of the additional phase conductor as well as because of strong inductive coupling between the phase conductor with current I_t on the one side and the additional phase conductor and all surrounding metal installations on the other side, and
- the length of HV cable lines between two adjacent HV substations supplied by the same cable line is such that almost always the following relation is satisfied:

$$Z_A \approx Z_B < Z'_{ph} L_S, \tag{6.9}$$

where L_S is the length of cable line section between transit and tested substation.

At the same time, this means that in practical conditions the influence of the length of the feeding line section used for a test circuit on test current distribution is also negligible.

6.3.2 Test circuit formed by using overhead section of feeding line

Overhead lines are also used in contemporary HV distribution networks, especially in suburban areas, that is, in areas with a lower degree of urbanization. Because of this, when the feeding line section between the tested and transit substations is an overhead line, the test current underground distribution should also be considered from the standpoint of the interfering inductive influence of metal installations surrounding the test circuit. Here, it is important to point out that overhead HV lines in distribution networks are relatively short and because of that they are usually with no transposed phase conductors.

As well as in the previous case, in this case for forming the necessary test circuit, we will select two phase conductors of the used feeding line section. Since the analogy with the previous case is obvious, the same line model and the same equivalent circuit can be used as well as in the previous case (Figs. 6.2 and 6.3). Also, the necessary analytical expressions for impedances Z_{ph}, Z_{01}, Z_{0eq}, and Z_{1eq} are the same, only notation d has a somewhat changed meaning. In this case, it denotes the distance between two used phase conductors of the overhead line.

In this case, we assume that the equivalent conductor is also a cylinder and that its center is in the middle of the fictitious line connecting the phase conductor with current I_t and the phase conductor with current I_1. Also, the ground wire(s) of the used overhead section should be disconnected from one of the grounding electrodes at this section ends. Under such conditions the relevant parameters of the introduced equivalent conductor can be obtained by using a calculation procedure analogous to the one in the previously considered case.

6.3.3 Line section composed of one overhead and one cable subsection

This case very rare in the contemporary power distribution practice is solved in Refs. [20,21] and considered in Chapter 4, Reduction factor of lines consisting of one overhead and one cable section. Since in this case as well as in the previous cases two phase conductors are necessary for obtaining a test circuit with currents I_t and I_1, the analogy with the previous cases is obvious. The only difference emanates from the fact that the mutual distance between the used phase conductors is different for different parts of such a feeding line section. Because of that the mutual impedance between the used phase conductors Z_{01} is determined by the following relation:

$$Z_{01} = L_c Z'_{01c} + L_o Z'_{01o} \tag{6.10}$$

where L_c (L_o) is the length of cable (overhead) part of the used feeding line section (km), and Z'_{01c} (Z'_{01o}) is the longitudinal mutual impedance between the used phase conductors of the cable (overhead) part of the used feeding line section (Ω/km).

Here, it can be mentioned the case of HV lines having also sections with the mutually different relevant parameters because of the special measures undertaken with the aim for improving their reduction factor. These cases are separately considered in Chapter 4, Reduction factor of lines consisting of one overhead and one cable section and also become solvable by the application of the here-presented methodological version.

6.3.4 Method applicability

As is shown in all considered cases for obtaining current I_1 instead of one neutral conductor (cable sheath or ground wire) used in the methodology version in Ref. [3], one additional phase conductor is used. Since the conductivity of the phase conductors is significantly larger than the conductivity of the neutral conductors, the error in the analytical part of the method introduced by the approximations, $Z_A \approx Z_B \approx 0$, is significantly smaller.

The here-presented version of the methodology of compensation of deficiency of relevant data, or Popović's methodology [18], enables taking into account the influence of all surrounding metal installations regardless of the complexity of the electrical circuit formed around and along the used line section and regardless of the number of unknown data. Some inaccuracies can appear only as a consequence of the inductive influence of the nearby power distribution lines. This influence can be efficiently avoided by using the test current with a frequency somewhat higher than 50 or 60 Hz that can easily be discriminated from the omnipresent power frequency (e.g., [22]). The error introduced in this way is small and gives the final results that are slightly on the side of enlarged safety.

The procedure analogous can be used for obtaining the actual reduction factor of a feeding line. It is necessary only that in that case the test circuit involves the whole feeding line and that the neutral conductors of the feeding line are included in the test circuit (disconnection D in Fig. 6.1 should be in the closed position) as shown in Chapter 3, Actual reduction factor of power distribution lines.

6.4 Actual grounding system impedance of a tested substation

When the values of the relevant parameters of the equivalent conductor (R'_{eq} and r_{eq}) are determined, the impedances Z_{eq}, Z_{0eq}, and Z_{1eq} can be determined by using Eqs. (6.5) and (6.6). Then, by using the equivalent circuit shown in Fig. 6.2 and Kirchhoff's Rules, one can determine the required current injected into the ground, I_e. However, due to the high conductivity of phase conductors and strong inductive coupling between them, current I_e obtained by the previous test circuit (Fig. 6.1) is too small for creating a measurable (sufficiently high) potential on the tested grounding system. Because of that one additional test circuit should be formed for each of the previously considered cases. This circuit is formed by omitting the previously used additional phase conductor so that the modified test circuit becomes as shown in Fig. 6.4.

The used notation has the following meaning: V is the voltmeter.

Based on known impedances Z'_{eq} and Z'_{0eq}, we obtain that the reduction of current I_e solely because of the metal installations surrounding the test circuit is the following:

$$r_t = 1 - \frac{Z'_{0eq}}{Z'_{eq}} = \frac{R'_{eq}}{R'_{eq} + \omega_t \frac{\mu_0}{8} + j\omega_t \frac{\mu_0}{2\pi} \ln \frac{\delta_t}{r_{eq}}} \tag{6.11}$$

Then, by using measured I_t and V_t, we obtain

$$Z_B = \frac{V_t}{I_e} = \frac{V_t}{r_t I_t}, \tag{6.12}$$

where V_t is the potential of the grounding electrode in substation B (V).

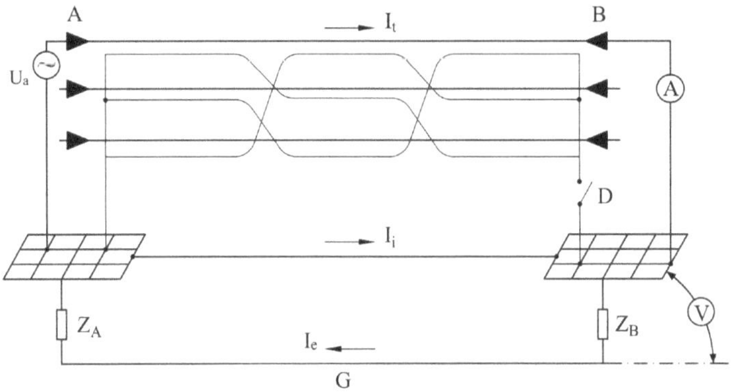

Figure 6.4 Additional test current circuit.

With the aim to eliminate the mutual influences between the potential measurement circuit and metal installations surrounding this circuit, this potential is, according to Ref. [4], determined by

$$V_t = k_1 k_2 V_m, \tag{6.13}$$

where V_m is the grounding electrode potential obtained by measurement (V); k_1 is the voltmeter correction factor; and k_2 is the relative position correction factor.

A voltmeter correction factor is determined by

$$k_1 = \frac{R_V + (Z_p + R_E)}{R_V} \tag{6.14}$$

where R_V is the internal resistance of the voltmeter (nameplate value); $Z_p + R_E$ is the impedance of the potential measurement circuit seen from the voltmeter terminals (this value is measured by connecting a Megger directly to the voltmeter terminals); and R_E is the grounding resistance of the potential electrode.

Relative position correction factor k_2 is determined by

$$k_2 = \frac{1}{(p/d_a) + (p/d_b) + (p/d_c) + 1} \tag{6.15}$$

where

$$p = \frac{1}{2}\sqrt{\frac{A_B}{\pi}}, \quad \text{m}$$

A_B is the area of the substation under measurement (m^2); d_a is the distance between the current and potential electrodes (m); d_b are distances between the potential electrode and the limits of the tested grounding electrode (m); and d_c is the distance between the current electrode and the limits of the tested grounding electrode (m).

The value of the grounding resistance R_E should be kept small, to improve the quality of the measurements. However, when an electronic voltmeter of high internal resistance ($>10^7\ \Omega$) is used, this requirement is of secondary importance. Also, the direction of the LV cable used for the potential measurement should be as perpendicular as possible to the current test circuit to minimize interferences.

The value of current I_e, according to the former practice, was obtained as a ratio between the tested grounding system (electrode) potential and the difference of the measured test current and the measured sum of currents circulating through the cable sheaths during the measurements. However, in this way, according to Fig. 6.4, the currents induced in the surrounding metal installations are completely ignored. Because of that the grounding system impedances obtained in this way were just apparent ones that can be several times smaller than the actual ones as is shown in Section 6.6.

In highly urbanized areas the actual reduction factor of the feeding line section used for the test circuit can be significantly smaller than the actual reduction factor of the whole feeding line. In such cases the influence of the metal installations surrounding the feeding line section applied for the test circuit can significantly be stronger than the influence of the metal installations surrounding the rest of the feeding line [3,21]. Thus without using the presented procedure, the safety conditions of the tested substations are assessed as more favorable than they are in reality. Due to this in such cases, it is quite possible to conclude that the prescribed safety criteria are satisfied, although this is not the case. The possibility of such an inadmissible error exists in all HV stations located in urban areas and supplied via at least one transit substation. To eliminate the possibility of such an error when testing the safety conditions, it is necessary to determine the actual ground impedance of these substations. Based on this impedance, preliminary estimation of safety conditions can be performed and reliable conclusions can be obtained concerning the question, whether they have been met in the tested HV substation.

6.5 Preliminary estimation of the tested grounding system

6.5.1 *Procedure description*

On the basis of the determined value of the actual grounding system impedance and the value of the anticipated maximal ground-fault current in the tested substation, the potential on the grounding electrode of this substation can be estimated with some reserve on the safe side by using the following expression:

$$V_B = r_c Z_B I_f \tag{6.16}$$

whereas the potential differences (touch and step voltages) essential for the estimation of the safety conditions can be determined by

$$U_m = U_{mt} \frac{I_f}{I_t}, \tag{6.17}$$

where I_f is the anticipated largest value of the ground-fault current passing through the grounded neutral points in the supply network (A); r_c is the feeding line reduction factor obtained by calculation; U_m is the maximal touch voltage that can appear on the grounding system (V); and U_{mt} is the maximal touch voltage obtained by test measurements (V).

The calculated reduction factor of a cable line with applied cross-bonding is determined by Eqs. (6.16) and (6.17). The necessary analytical expressions in the case of HV overhead lines are presented in corresponding technical standards, for example, [2], whereas if a feeding line is a combination of different sections, these expressions are developed in Refs. [20,23] and presented in Chapter 4, Reduction factor of lines consisting of one overhead and one cable section and Chapter 5, Measures for improving reduction factor of a feeding line.

Since we are dealing here with a problem that refers exclusively to HV substations located in urban and suburban areas, here is necessary to mention that one of the specificities of their grounding systems is of special importance. That is a fact that as the most critical from the point of view of safety conditions on the whole grounding systems of these substations is, as a rule, the potential transferred through the metal sheaths of outgoing MV cable line to the nearest supplied MV/LV substation [6]. This potential is determined by

$$V_{tr} = k_{tr} V_B, \tag{6.18}$$

where k_{tr} is the coefficient expressing the relative value of the potential transferred to the nearest MV/LV substation.

In the case of MV cable lines having insulated metal sheaths, the most common MV cable lines in contemporary power distribution network, it is not difficult to show that coefficient k_{tr} is determined by

$$k_t = 1 - \frac{l_s Z'_{3sh}}{Z_S}, \tag{6.19}$$

where l_s is the length of the outgoing cable line up to the nearest MV/LV substation, (m); Z'_{3sh} is the self-impedance of the equivalent sheath substituting the sheaths of the three MV single-core cables per one meter (Ω/m); and Z_S is the grounding impedance of the nearest MV/LV substation, excluding the grounding effect through this cable line as a conductive connection with the grounding system of the tested HV substation (Ω).

Impedance Z_{3sh} is, according to Eq. (2.20) in Chapter 2, determined by

$$Z'_{3sh} = \frac{R'_{sh}}{3} + \omega_t \frac{\mu_0}{8} + j\omega_t \frac{\mu_0}{2\pi} \ln \frac{\delta_t}{\sqrt[3]{2 r_{sh} d^2}}, \quad \Omega/\text{km}, \tag{6.20}$$

whereas potential V_{tr} can be obtained by measurement during the here-described test measurements (Fig. 6.4).

However, because of the inductive influence of the metal installations surrounding a feeding line, the actual value of the reduction factor can significantly be smaller than the one obtained by calculations. Due to this, the previously described estimation procedure gives results that are excessively on the conservative side in the case of HV substations located in urban and suburban areas.

In the cases when a feeding line has the same constructive characteristics along whole its length and when the results of the previously described preliminary estimation are not satisfactory, a somewhat more accurate preliminary estimation can be made if the section used for the test circuit is in an area with a greater degree of urbanization than the rest of the feeding line. Then instead of the calculated reduction factor of the foreseen feeding line, one can use the actual reduction factor of the rest of the feeding line but determined for the adjacent transit substation. Certainly, this reduction factor can also be used at the design stage of such HV substations, since it is lower (more favorable) and nearer to the actual reduction factor of the future feeding line than its calculated reduction factor.

In critical cases, when the preliminarily estimated maximal transferred potential again exceeds prescribed (tolerable) limits, it is necessary to perform one more estimation but this time by using the actual value of the reduction factor of the feeding line. However, this implies taking into account the inductive influence of all surrounding metal installations existing in the conditions of an actual ground-fault. It is possible only when the test circuit is formed in such a way that the test current circulates along the entire feeding line length.

Finally, HV distribution substations usually have more than one connected HV line that can be used as a feeding line in certain regular or irregular circumstances in a supply HV network. The maximum (critical) earth current in such cases cannot be known in advance, that is, only on the basis of the largest value of the anticipated ground-fault current. The largest value of so-called earth current can be in cases where anticipated ground-fault current, I_f, is not the largest one. This can be encountered in cases where the feeding line causing maximum ground-fault current passes through the area that is with the most degree of urbanization. Due to this, in some cases it is necessary to perform test measurements for each of the possible feeding lines, and, on the basis of comparing the different values for earth current, determine the largest earth current, critical (worst) for safety conditions.

The flowchart describing the whole here-described process of the preliminary estimation of safety conditions within and in the vicinity of an HV is shown in Fig. 6.5.

The used notation has the following meaning: V_t is the potential of the grounding electrode of the tested substation obtained by measurements; V_p is the permissible potential transferred to the nearest MV/LV substation; r_{Aa} (r_{Ba}) is the actual reduction factor of the feeding line supplying the transit substation, A (B); and Δt is the duration of ground-fault current.

Potential V_p can be somewhat higher than the maximal permissible touch voltage depending on the type of TN earthing system implemented in the LV network [14].

In urban and suburban areas, many metal installations act as perfect external grounding electrodes of HV substations located in urban areas, as well as additional

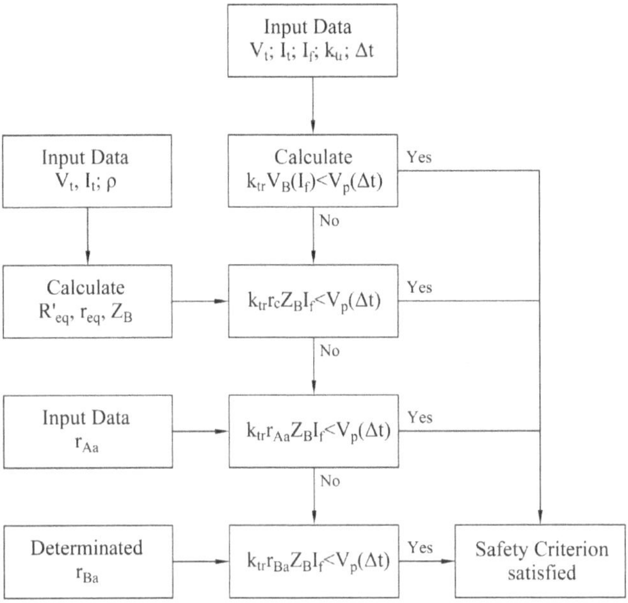

Figure 6.5 Process diagram of the preliminary estimation.

neutral conductors of their feeding lines [6,12]. Thus it is quite realistic to assume that the preliminary test will be sufficient for definitely obtaining a favorable assessment of safety conditions of many of the HV substations tested in here-described way.

Certainly, in cases where there is no transit substation between the tested and the supply substation, that is, where the entire feeding line is used for the test circuit, the test measurements are performed in conditions that do not differ from those that will exist in the case of an actual ground-fault. Then it is sufficient for a correct assessment of the safety conditions to measure V_t and to multiply its value by coefficient k_{tr} and the ratio of the largest ground-fault current and the test current, I_f/I_t. The same procedure can be used for cases in which a section of the feeding line passing through an area with an evidently lower degree of urbanization than the rest of the line is used to form the test circuit. In such cases the problem arises only if the results of this test do not show that the prescribed safety criteria are met. Then the test measurements must be repeated, but this time in such a way that the formed test circuit uses the entire feeding line, from the supply to the tested substation.

6.5.2 A practical example

Experimental investigations on the inductive influence of metal surrounding installations are performed on a cable line that supplies, in series, two substations in the 110-kV distribution network of Beograd, Serbia [3,8]. The length of the line to the closer of the supplied substations measured from the supply substation is 2320 m, whereas the total feeding line length to the tested substation is 6590 m. The line is realized by single-core cables, laid in a trefoil formation on the entire length. The cross-bonding technique used for the elimination of currents circulating through the cable sheaths was not applied to this feeding line.

In the areas through which the line passes, the equivalent soil resistivity is estimated to be in the range of from 30 to 50 Ω m, and as a less favorable from the standpoint of the value of reduction factor, the lower one has been used for necessary calculations. The line section between the source and the transit substation passes through an area with a lower degree of urbanization than the rest of the line. The phase conductors are made of aluminum with a cross section of 1000 mm^2, while the metallic sheaths are made of copper strings with a total cross section of 95 mm^2 and a mean diameter that is equal to 91 mm. The main elements of the grounding system in both of the supplied substations are the substation building foundation and the 44 outgoing cable lines realized by an armored lead sheathed three-core cable.

In the first attempt the test circuit was formed by using only a feeding line section between the tested and the transit substations of the length of 4270 m. The second test circuit was formed by using the whole feeding line. While measurements were being taken in both of these lines, only the metal sheath of the cable with the test current, I_t, was grounded at both of the test circuit ends.

On the basis of the measurements of the test current through one of the single-core cables and current induced in its sheath, the following results have been obtained:

$I_1/I_t = 0.7445 + j0.3790$, for the shorter test circuit,
$I_1/I_t = 0.7585 + j0.3623$, for the longer test circuit.

With the aim of enlarging the ground current, I_e, and obtaining a measurable value of the potential of the ground electrode of the tested substation, an additional measurement was performed. This measurement was made by disconnecting all feeding line metal sheaths at one of the test circuit ends. Then, for the test current and potential at the grounding electrode of the tested substation were obtained the following values:

$I_t = 136$ A and $V_t = 1.14$ V, for shorter test circuit;
$I_t = 85$ A and $V_t = 1.20$ V, for longer test circuit.

By using Eq. (6.8) and these results, it follows that the relevant parameters of the equivalent sheaths, substituting all surrounding metal installations, are the following:

$R'_{eq} = 0.031$ Ω/km and $r_{eq} = 7.1$ m, for the shorter test circuit;
$R'_{eq} = 0.053$ Ω/km and $r_{eq} = 8.1$ m, for the longer test circuit.

By using Eq. (6.11) and results obtained for R'_{eq} and r_{eq}, the values of the earth current are

$I_e = (4.23 - j14.15)\ A$, for the shorter test circuit;
$I_e = (5.89 - j15.00)\ A$, for the longer test circuit.

When these currents are determined, on the basis of the measured value of corresponding potential V_t, the identical value for the substation grounding impedance is obtained for both of the considered cases. The value of this impedance is

$$Z_B \approx 0.075\ \Omega.$$

Certainly, the value of the actual grounding system impedance can have only one value, whereas the apparent grounding system impedance is different for different surrounding metal installations. In the cases when the interfering influence of the metal installations surrounding the test circuit is not eliminated, this impedance would be, according to Eqs. (6.11) and (6.12), considered 81.0% smaller for the longer and 88.9% smaller for the shorter test circuit. Before the investigation results presented in Ref. [8], these apparent values were treated as actual ones. This is probably the main explanation for the fact that the measured values of the grounding system impedance of all 110/10 kV substations in the distribution network of Beograd were, almost without an exception, several times smaller than the ones obtained by calculations at the design stage.

When the actual grounding system impedance has been determined for the first preliminary estimation of the safety conditions within and in the vicinity of the tested HV substation, it is sufficient to know the value of the calculated reduction factor of the feeding line. By using Eq. (6.16) the calculated reduction factor in the case of the considered line is

$$r_c = 0.020 - j0.103$$

However, for obtaining an actual picture concerning the safety conditions, it is necessary to determine the actual reduction factors in both of the experimentally obtained cases for the feeding line. Using the methodological version presented in Refs. [3,8] has shown that the relevant data of the equivalent sheath substituting only surrounding metal installations are the following:

- from the tested to the transit substation:

$$R'_{eq} = 0.031\ \Omega/\text{km} \quad \text{and} \quad r_{eq} = 7.1\ \text{m}$$

- from the transit to the source substation:

$$R'_{eq} = 0.08\ \Omega/\text{km} \quad \text{and} \quad r_{eq} = 12.3\ \text{m};$$

whereas for the whole feeding line, they are:

$$R'_{eq} = 0.053\ \Omega/\text{km} \quad \text{and} \quad r_{eq} = 8.1\ \text{m}$$

Based on these results and the relevant data concerning the cable line sheaths, the actual reduction factor of these line sections can be determined by using, according to [8], the following expression:

$$r_a = \frac{R'_{3sh}R'_{eq}}{Z'_{3sh}Z'_{eq} - Z'^{2}_{3sheq}}, \tag{6.21}$$

where Z'_{3sheq} is the self-impedance of the equivalent sheath substituting the sheaths of the three single-core cables per one meter (Ω/km), and Z'_{3sh} is the mutual impedance between the equivalent sheath substituting the sheaths of the three single-core cables and the equivalent sheath substituting all surrounding metal installations per one meter (Ω/km).

In the case of cables laid in trefoil formation impedance Z'_{3sh} is, according to Eq. (6.20), determined by

$$Z'_{3sh} = \frac{R'_{sh}}{3} + \omega_t \frac{\mu_0}{8} + j\omega_t \frac{\mu_0}{2\pi} \ln \frac{\delta_t}{\sqrt[3]{4r_{eq}}}, \quad \Omega/\text{km}, \tag{6.22}$$

whereas impedance Z'_{3sheq} is determined by

$$Z'_{3sheq} = \omega_t \frac{\mu_0}{8} + j\omega_t \frac{\mu_0}{2\pi} \ln \frac{\delta_t}{\sqrt[3]{4r_{eq}}}, \quad \Omega/\text{km}, \tag{6.23}$$

The obtained results of performed calculations show that the values of actual reduction factor for the transit substation, r_{AB}, the section between source and transit substations, r_{Aa}, and the entire feeding line, r_{Ba}, are the following:

$$r_{AB} = -0.018 - j0.012; \quad r_{Aa} = -0.031 - j0.037; \quad r_{\text{Ba}} = -0.026 - j0.023.$$

By using these values and the following also relevant data: $k_{tr} = 0.8$ and $U_p(\Delta t = 0.5\ \text{s}) = 164$ V, the first preliminary assessment shows that the prescribed safety criteria are met if the ground-fault current is less than 20.8 kA. For the same data and the value of r_{Aa}, the prescribed safety criteria are met if this current is less than 45.3 kA. In the case when the test circuit has been formed over the entire feeding line length, the prescribed safety criteria are met if this current is less than 63.0 kA.

As can be seen that the actual reduction factor of the section used for the test circuit, r_{AB}, is significantly lower than the actual reduction factor of the feeding line. Thus the test circuit formed by using the transit substation grounding system as an auxiliary current electrode, in this case, gives a more favorable assessment of the safety conditions than they really are. Because of that, in these cases, such a test circuit and the obtained results can be used only for obtaining the actual grounding system impedance and preliminary estimation of the safety conditions. Since the largest ground-fault current in the considered case was 20 kA, a favorable assessment concerning the safety conditions was obtained already after the first attempt foreseen by the presented iterative procedure for preliminary assessment.

As it has been shown in Chapter 2, Calculated reduction factor of feeding cable lines, the relative distribution of the ground-fault current between individual sheaths is somewhat changed when we apply cross-bonding on a certain power cable line. However, based on the identical analytical expressions for the reduction factors of power cable lines with no cross-bonding and the cable lines but with cross-bonding, it is quite realistic to assume that the inductive influence of the surrounding metal installations cannot be noticeably changed by the cross-bonding application. The possible differences are especially negligible in cases when the mutual distance between single-core cables of one HV or EHV line is relatively small in relation to their distance from the surrounding metal installations. Since in the considered case the single-core cables touch each other, we can consider that the presented results of the experimental analysis are identical to the results that would be obtained for the same lines but with applied cross-bonding.

The presented results relate to only two different cases from the standpoint of the feeding line length and the surrounding metal installations (the feeding line to the tested substation and the feeding line to the transit substation). However, the given example is very illustrative and shows that the actual distribution of the ground-fault current can be correctly determined only if the test circuit is formed by using the entire feeding line.

6.6 General observations

When the section used for obtaining the test circuit is passing through an area of a higher degree of urbanization than the rest of the feeding line, the safety conditions evaluated without using the here-presented test procedure can be seen as more favorable than really are. According to the presented practical example, the grounding potential rise obtained on the basis of the earlier known test procedure is about 37.7% lower than the actual one. The difference can be even significantly larger in the cases when the rest of the feeding line is constructed as an overhead line as is the case considered in Ref. [21]. In such cases, if the main factors influencing the safety conditions (high specific soil resistivity and/or high level of ground-fault current) are very unfavorable, without eliminating the interfering inductive influence of metal installations surrounding a test circuit, the definite assessment about the safety conditions within and in the vicinity of some HV distribution substations can be completely wrong. In other words, the prescribed safety criteria can be considered satisfied, although this is not the case and too high transferred potentials can appear at certain, exposed places of the large urban area around such substations.

The fraction of the ground-fault current dissipated through the grounding system into the surrounding earth in the cases of HV substations located in the urban area of an average degree of urbanization is about two to four times smaller than it was previously considered. This research result throws completely new light on the entire grounding problem of HV distribution substations and dramatically changes our earlier perception about the magnitude of this problem. Now, this problem can be seen in the realistic frameworks enabling, generally speaking, solutions that are in accordance with the real needs, that is, without redundant expenditures.

Not to remain only on the given practical example, one can assume the cases where the urban environment is the same, but only the equivalent soil resistivity is gradually increased that can be in reality, as is well known, a consequence of a long-lasting dry period. Then, as well as in the case of the current through feeding line neutral conductors, the currents through the surrounding metal installations become greater and greater, while the earth current becomes smaller and smaller. According to this observation, the final calculation results will be slightly on the safe side if we for the calculations necessary for obtaining R'_{eq} and r_{eq} adopt a value of the equivalent soil resistivity that is approximately estimated as the lowest possible for the urban area along and around a considered feeding line. The reduction of the earth current or current dissipating through the grounding system of the supplied substation into the surrounding earth is also greater if the urban environment is on a greater degree of urbanization. Finally, it should be mentioned that due to the greater mutual distance of the neutral conductor(s) from the phase conductors, the relative participation of the surrounding metal installations in the distribution of the ground-fault current is larger in the case of overhead lines.

Also, on the basis of the obtained results, it is clear that a significant part of the test current passes through the different metal installations surrounding the feeding line. This means that under actual ground-fault conditions, the current passing through a certain surrounding metal installation can be several kilo Amperes. Consequently, from the standpoint of public safety, it is important to stress that all these metal installations should be mutually conductively connected. This requirement, as is well known, should be already fulfilled through the strict and consistent application of TN earthing systems in LV networks and customers' installations [13].

The test method presented here enables a preliminary estimation and in many cases definite judgment concerning the safety conditions within and in the vicinity of HV/MV substations supplied via one or more transit substations. The whole procedure is based on determining the actual impedance of the grounding system of this HV substation. The necessary elimination of the interfering inductive influence of the metal installations surrounding the test circuit is achieved by the application of the here-presented methodology.

Having in mind the very low values of grounding system impedances of HV distribution substations, for example, [6,12], it is quite realistic to assume that the preliminary test procedure presented here will not be sufficient for obtaining a definite assessment concerning the safety conditions in a very small number of cases. In such cases, which can occur due to high ground-fault current and/or too high local soil resistivity, testing should be performed in conditions that are the same as those that exist in the case of an actual ground-fault. This means that the test circuit should be formed in such a manner that the test current circulates along the entire feeding line, that is, from the source to the tested substation.

The presented test procedure enables the determination of the actual grounding system impedance of an HV substation by eliminating the interfering inductive influence of metal installations surrounding the test circuit in urban surroundings. By using this procedure, it is possible to obtain a reliable preliminary assessment of the safety conditions at periodical testing and verification of the safety conditions for HV substations located in urban or suburban areas and supplied via one or more transit substations. By ignoring the interfering inductive influence of surrounding metal installations, the definite judgment concerning the safety conditions in some cases of HV distribution substations can completely be wrong.

References

[1] IEEE, Guide for Safety in Substations Grounding, Std. 80−2000.

[2] Int. Std., ref. CEI/IEC 60908−3, (Ed.) 0.3 2009−3, Short-Circuits Currents in Three-Phase A.C. Systems—Part 3: Currents During Two Separate Simultaneous Line-to-Earth Short Circuits and Partial Short−Circuits Following Through Earth.

[3] L.M. Popović, Testing and evaluating grounding systems for substations located in urban areas, IET generation, Transmission and Distribution 5 (2) (2011) 231−238.

[4] P.R. Pilai, E.P. Dick, A review on testing and evaluating substation grounding systems, IEEE Transactions on Power Delivery 7 (1) (1992) 53−61.

[5] S.T. Sobral, J.J. Horta Santos, D. Mukhedkar, A proposal for ground measurement techniques in substations fed exclusively by power cables, IEEE Transactions on Power Delivery 3 (4) (1988) 1403−1409.

[6] L.M. Popović, Comparative analysis of grounding systems formed by MV cable lines with either uninsulated or insulated metal sheath(s), Electric Power System Research 81 (2) (2011) 393−399.

[7] L.M. Popović, Active length of cable lines with uncoated metal sheaths as long grounding conductors, ETEP European Transaction on Electrical Power 22 (2) (2012) 206−215.

[8] L.M. Popović, Influence of the metallic installations surrounding the feeding cable line on the ground fault current distribution in supplied substations, IEEE Transactions on Power Delivery 23 (4) (2008) 2583−2590.

[9] M. Lucheroni, Measurements of touch and step voltages adopting current auxiliary electrodes at reduced distance, IEEE Transactions on Industry Applications 44 (6) (2008) 1896−1901.

[10] G. Parise, L. Martiraneo, L. Parise, S. Gelozzi, R. Araneo, Simplified conservative testing method of touch and step voltages by multiple auxiliary electrodes at reduced distances, IEEE Transactions on Industry Applications 51 (6) (2015) 4987−4993.

[11] IEEE Guide for Bonding Shields and Sheaths of Single-Conductor Power Cables Rated 5 kV through 500 kV, IEEE Standard 575−2014, 2014.

[12] L.M. Popović, Practical Methods for Analysis and Design of HV Installation Grounding Systems, Academic Press, Elsevier, Oxford, 2018.

[13] Int. Std., CEI IEC 61936−1 Power installations exceeding 1kV a.c. Part 1: Common rules, First edition 2002−10.

[14] Int. Std., IEC 60364−1 Low-Voltage Electrical Installations-Part 1: Fundamental Principles, Assessment of General Characteristics, 2005 Edition.

[15] L.M. Popović, Ground-fault current distribution when a ground fault occurs in HV substations located in urban areas, Progress in Electromagnetic Research B 5 (2014) 167−179.

[16] L.M. Popović, Inductive influence of HV cable lines in urban and suburban areas, Electrical Power System Research 176 (2019). Available from: https://doi.org/10.1016/j.epsr.2019.105944.

[17] L.M. Popović, Inductive influence of surrounding metal installations on power transfer capacity of HV cable lines with applied cross-bonding, International Journal of Electrical Power & Energy Systems 120 (2020) 105963.

[18] L.M. Popović, Preliminary testing and assessment of safety conditions of HV substations located in urban areas, Electric Power System Research 195 (2021) 107111. Available from: https://doi.org/10.1016/j.epsr.2021.107111.

[19] R. Benato, et al., Ground return behavior in high voltage alternating current insulated cables, Energies 7 (12) (2014) 8116–8131.

[20] L.M. Popović, Ground fault current distribution in a station supplied by a line composed of a cable an overhead section, ETEP, European Transactions on Electrical Power 17 (3) (2007) 207–208.

[21] J. Wilas, D. Mukhedkar, V. Fernandes, A. Magalhaes, Grounding grid design of a transition station—a typical example of fault transfer, IEEE Transactions on Power Delivery 5 (1) (1989) 124–129.

[22] J. Waes, M. Riet, F. Proroost, S. Cobben, Measurements of the current distribution near a substation during a single phase to ground fault, in: Proceedings of CIRED Conference, 2003.

[23] L.M. Popović, Reduction of the fault current passing through the grounding system of an HV substation supplied by cable line, International Journal of Electrical Power & Energy Systems 99 (0) (2018) 493–499.

Inductive influence of high-voltage and extra-high voltage lines on surrounding metal installations

7

7.1 Inductive influence on the nearby metal installation

7.1.1 Introduction

During a ground fault the possibilities of appearances of dangerous and harmful voltages and currents are not limited only to the areas within and in the vicinity of high-voltage (HV) substations located in an urban or suburban area. As a consequence of the inductive influence of its feeding line on the surrounding metal installations, these voltages and currents can also appear at places along this line that are very far from the substation having a ground fault. Because of that, for the designers of HV distribution substations, it is necessary to have on a disposal a sufficiently correct methodology enabling the determination of the voltage and current that can be induced in any of the metal installations surrounding feeding lines of these substations.

The electromagnetic interaction between an HV or extra-high voltage (EHV) line and a neighboring metal installation can be divided into the following three categories: capacitive (electrostatic), conductive (resistive), and inductive (magnetic) coupling. The first one is generated by the electric field, which influences only metal installations above the ground, having no path to the ground. Owing to the first interaction, a metal installation picks up a voltage relative to the soil that is proportional to the power line voltage. The second one occurs during lightning strikes or at the phase to ground faults. When this has occurred, a large voltage cone is created around the grounding electrode and as a final consequence, the too high voltage can penetrate the surrounding metal installation through the installation's coating defects. The third one is inductive influence, which is generated by the fluctuating magnetic field present during both normal operating conditions and fault conditions when the nearby installation is placed in parallel or almost parallel to an HV or EHV line. Here, we will deal only with the inductive influence of HV and EHV lines when they have been constructed in an urban environment.

In the urban environment, because of many mutually inductively coupled metal installations typical for urban areas, the inductive influence appears as a very complex phenomenon in the creation of which each of the metal installations affected by the magnetic field has a certain participation. This is probably one of the reasons

Relevant Characteristics of Power Lines Passing through Urban Areas. DOI: https://doi.org/10.1016/B978-0-323-91136-8.00008-6

because of which, this problem was until recently without an adequate methodology for its solving, although it was known to the wider professional publicity practically from the very beginning of massive electrification [1–6].

When a communication line or a metal pipeline runs parallel to a power line along its whole length or only along some of its sections, certain voltages are induced in this installation under both normal operation and fault conditions in this power line. These voltages can be so high that they represent an electrical shock hazard for the operating personnel or the general public. They can be also a threat to the integrity of cathodic protection equipment, pipeline coating, or pipeline steel, particularly during ground-fault conditions.

According to the international regulation, practice potentials above 15 V should be considered hazardous and measures should be undertaken toward reducing these potentials. Therefore the inductive influence of HV or EHV lines represents an electromagnetic phenomenon that is the subject of investigations and consideration in all practically possible conditions and circumstances in which it can occur and cause unwanted consequences. This influence may be especially intensive in the case of a ground fault within a directly earthed HV network and the case of a long section of the exposure of the induction-affected installation.

Because of all this, the prevention of dangerous and harmful consequences of the inductive influence of HV and EHV lines on nearby insulated conductors (communication lines or different metal pipelines) has been a subject of research for almost the last 100 years. As a result, many of so far published papers have proposed a relatively large number of methodologies for solving this problem in different conditions and circumstances. A short review of their development, improvement, and promotion through this relatively long research period can be summarized in the following manner.

The first attempts to study the AC interferences are based on the widely known Carson's equations [7], separately considered and presented in Chapter 1, Introductory considerations. Somewhat later, Pollaczek described in Ref. [8] inductive coupling between parallel conductors in the presence of earth as a half-space conductive medium. Paper in Ref. [9] extended Carson's work to multilayer earth and conductors near point sources of current. A later paper in Ref. [10] included the electrical characteristics of pipelines such as resistance to earth of a bitumen-coated pipeline surface.

In 1986 Dawalibi and Pinho [11] described computer-based analysis aimed at ensuring personnel safety and pipeline integrity during power system faults. Then, Sobral et al. [12] applied the Decoupled Method to simplify the calculation of the interferences affecting a single communication circuit or a single pipeline, caused by a ground fault in a single nearby power transmission line.

Also, paper in Ref. [13] demonstrated, by using computer simulation software, the performance of a highly effective AC interference mitigation method. The method, which combines the effectiveness of grounding conductors and gradient control wires, attenuated both inductive and conductive interference providing, at the same time, cathodic protection. Then a method for the calculation of currents and voltages in a system of parallel pipelines and buried the conductors was proposed in Ref. [14].

The method is based on the lossy transmission line model for buried conductors and the previously mentioned methods using the same approach but modified by abandoning the Thevenin circuit representation of the pipeline sections.

The more recently published paper in Ref. [15] presented a hybrid method employing finite element calculations and standard circuit analysis to calculate the induced voltages and currents in a pipeline running parallel to a faulty transmission line. Then, the paper in Ref. [16] developed a simulation software tool for prediction and mitigation techniques applicable to pipeline networks influenced by HV power lines. The software that can deal with any configuration (no limitation on the number of pipes, transmission lines, bonds, groundings, coating, and soil resistivity) is very user friendly and robust since a general algorithm is used to calculate the EMF (induced electromagnetic force), eliminating the need for subdivision of pipelines in sections, parallel or not, in relation to the transmission line(s). With the calculated values of EMF, the induced voltages and currents are then obtained by solving the well-known transmission-line model using a numerical technique that specifies the pipeline parameters (diameter, soil resistivity, etc.) for each section of the pipeline. Finally, a rigorous solution for the electromagnetic field of underground conductors and two-layer earth is presented in Refs. [17,18].

On the basis of the presented short review, one can note that considerable efforts have been made over the several decades for the development of sufficiently accurate computation models for calculating voltages and currents induced by power lines in the neighboring metal installations. Based on the developed mathematical models of actual physical phenomenon and computation techniques, several software packages are available nowadays. As an example of their big possibilities, the method presented in Ref. [16] takes into account an arbitrary number of pipelines, power lines, bonds, groundings, and coating at different soil resistivity.

The common characteristic of the nowadays methodologies is that they enable taking into consideration the relevant constructive characteristics and mutual spatial disposition of all surrounding metal installations. However, none of them enables solving the problem when it appears in urban and suburban areas, where many relevant data about the surrounding metal installations are uncertain and completely unknown.

As it is well known, the increasing sizes of modern power distribution networks, as well as the increasing operating and short-circuit currents of these networks, have been matched by the overspreading networks of other types of metal installations (different pipelines, different cable, overhead line neutrals, etc. closely laid along with HV and EHV distribution lines). Spatial dispositions of all of these installations belonging to a certain urban area are determined mainly by the spatial disposition of its streets, their length, and their winding. According to this, they are situated on a small mutual distance result in a strong inductive coupling and, in the vicinity of HV distribution substations, the conductive coupling between all of them. That is why the need for a methodology that enables solving problems in such conditions and circumstances has been present for a long time [19].

However, to solve the problem of inductive influence in such conditions, it was previously necessary to solve one more also important problem for the power distribution practice. This is the problem of ground-fault current distribution when a fault occurs in an HV substation located in an urban or suburban area. As is well known, in professional and scientific publicity, this problem was also a long time considered practically unsolvable. Not so long ago, after performed experimental measurements in the distribution HV network of Beograd, it has shown that the fraction of the ground-fault current flowing solely through the earth in the urban areas of average degree of urbanization is several (two to four) times smaller than it has known before [1]. This means that the great portion of the total ground-fault current is returning to its sources in the power system through the metal installations surrounding a feeding line. Somewhat later the method for determining actual ground-fault current distribution in the grounding system of an HV substation supplied by cable line with no applied cross-bonding has been improved and completed [2]. The developed methodology is based on the measuring of the currents through one phase and one neutral conductor of an HV line during the simulated ground fault in the supplied substation. The analytical part of the developed methodology is based on an introduced fictitious (equivalent) conductor substituting, from the standpoint of their inductive influence, all surrounding metal installations. It is imagined as a cylinder around all conductors of the certain HV or EHV line and along the entire length of this line. Also, the developed methodology is based on the analytical expressions for the self- and mutual impedances of line conductors installed in the semiinfinite and homogenous earth, for example [20].

After that it was shown that the introduced fictitious conductor can be determined in such a way that it substitutes all surrounding metal installations, including metal sheaths of the considered cable line [3,4]. The imagined physical appearance and spatial position of this conductor are such that it represents a cylinder along and around only the phase conductor carrying a simulated ground-fault current. As well as in the former case [2], this conductor is completely determined by only two relevant parameters. One of them is the mean radius of the cylinder representing this conductor, whereas the other one is the longitudinal resistance of this conductor. Somewhat later, by the same methodological framework, the first methodology enabling compensation of deficiency of many data relevant for the magnitude of the voltage induced in any one of the surrounding metal installations was developed [5]. This has been done by introducing into the developing procedure one equivalent conductor that can be, for the considered induced installation, treated as its equivalent screening conductor. It substitutes, from the standpoint of screening effects, all surrounding metal installations, including the inducing line neutral conductor (s). By introducing this equivalent screening conductor, it is possible to determine the current induced in any of the surrounding metal installations, provided that its technical characteristics and spatial position in relation to the considered inducing HV or EHV line are known. As a final result, the analytical expression for the resulting screening factor that involves the screening effect of all surrounding metal installations present along the section of exposure was derived. The equivalent screening conductor is imagined as a cylinder placed around the considered (mainly just at the design stage) installation only along its section of exposure. However, this method is not quite suitable for application because its application requires the accurate measuring of the potential induced at the end of a provisory use cable line. Since this is practically impossible in urban conditions, the further development of the methodology of compensation of deficiency of many relevant, but unknown data (or Popović's methodology), was necessary.

Instead of the previously used one phase and one neutral conductor for test measurements, this time two phase conductors are used. One of them during the simulated ground fault serves as an additional neutral conductor of the tested inducing power line. Unlike the previous version, forming of the test circuit and all necessary measurements are performed only within the substations at the ends of the inducing line. Also, it is shown that the mentioned fictitious (equivalent) conductor can be determined in such a manner that its physical appearance and spatial position coincide with any of the actual sheaths of the considered cable line. Thus the screening effect of all surrounding metal installations is expressed only through an apparent (expressed by a complex number) specific resistivity of the actual, from the standpoint of its physical appearance, sheath of the considered HV or EHV cable line. Finally, the developed methodology is such that it can be applied to any HV or EHV line, from the standpoint of its type and constructive characteristics as well as from the standpoint of conditions and circumstances existing in the area(s) in which it is constructed. The case of HV and EHV cable lines with applied cross-bonding, the most common case in contemporary power distribution networks [21], is also solvable by this methodological version.

Finally, for the application of the methodology version presented in Ref. [6], there is no need for collecting a large number of input data, it is sufficient relatively simple test measurements as well as relatively simple calculation procedures based on the results of those measurements. As such, easy to apply, it can be considered desirable also in cases that were solvable by using hitherto known methodologies. Its advantage upon these methodologies is especially pronounced in the case of problems that were so far solvable only by using modern computers.

7.1.2 Problem description

The large-scale power systems and pipeline systems, combined with the planning concept of concentrating the routes of various supply systems (electricity, water, gas, oil, telecommunications, etc.) have consequently led to a rise in interference problems. During a ground fault in the supplied substation, the voltages (longitudinal electromotive forces) induced by an HV or EHV line in surrounding metal installations can be so high that they represent an electrical shock hazard for the operating personnel or the general public. Besides, they are a threat to the integrity of the cathodic protection equipment, pipeline coating, or pipeline steel. Finally, the voltages and currents generated by the alternating magnetic field along and around an HV and EHV cable line can cause interference to neighboring communication and control systems.

Even under the unreal assumption that the phase voltages imposed at the beginning of a three-phase line are absolutely symmetrical, because of the different spatial position of the phase conductor(s) in relation to the earth and the neutral conductors, including surrounding metal installations, an unbalanced current always exists. Its value is certainly significantly smaller in comparison with a ground-fault current, but it flows permanently and because of that it can be also harmful and disturbing in some cases. This unbalanced current induces in a surrounding metal installation grounded at both ends the longitudinal electromotive force that causes a certain circulating current. This circulating current on its way to and from the earth passes through the grounding electrodes at the ends of the observed installation and thus creates a certain potential between this installation and the earth immediately outside the coating and thus can cause some damages. Also, as is well known, more sensitive electronic systems face an increasingly hostile electromagnetic environment. Magnetic coupling is an increasingly frequent cause of problems capable of upsetting data transmission. Thus in accordance with the corresponding technical regulations, the potentials in excess of 15 V should be considered hazardous and measures should be undertaken toward reducing them [22].

However, the currents induced in the surrounding metal installations should not be in all cases considered as dangerous and harmful and as such undesirable. In cases where we have more than one surrounding metal installation, the current induced in each of them reduces the voltages and currents induced in all other surrounding metal installations. Therefore with the increase in the number of surrounding metal installations, the problem of induced voltages and currents becomes smaller and smaller. From the point of view of the problem of the inductive influence of power lines, this favorable effect is known in the professional literature as the screening effect [19]. On the other side, with the increase in the number of surrounding metal installations, the complexity of the electrical circuit spontaneously formed by them around and along the inducing line also increases. Thus there is an increasing need for methodologies that can take into account a large number of elements of this complex electrical circuit in which the surrounding earth also participates as a specific alternating current conductor (Chapter 1, Introductory considerations). Because of all of this, the methodology sufficiently accurate for solving such an electrical circuit is especially desirable for the problems that appear in urban and suburban areas.

Certainly, in urban areas, the inductive influence appears under extremely complex circumstances because HV or EHV cable lines share the same routes with many other metal installations like metal sheaths of different types of other cable lines, neutral conductors of the low-voltage (LV) networks, steel water pipes, etc. Their basic functions are mutually different, but each of them spontaneously acts as an additional screening conductor and as such has a certain influence on the magnitude of induced voltages and currents in all other metal installations. Besides, many of them are laid under the surface of the street sidewalks and many relevant data concerning them cannot be visually or in some other way determined or verified. In other words, since many urban installations are invisible, many necessary data about them, constructive characteristics, mutual spatial disposition, including in some cases even and their total number, are uncertain or completely unknown. Only that is certainly known is the following:

their spatial disposition and structure depend on many local factors and, because of that, they are different from case to case;

some of them are in direct and continuous contact with the earth, whereas some others are grounded only at certain places, the most frequently in all buildings arranged along the streets; and

they are, through medium-voltage cable line metal sheaths, conductively connected to the grounding electrodes at the ends of an HV or EHV line, by strictly and consequently applied terra-neutral grounding system in the LV network and consumer installations.

By disregarding for a moment, all other surrounding metal installations, let us assume an HV cable line having one section running along the same street(s) as one planed metal installation. Such a situation is schematically depicted in Fig. 7.1.

The used notation has the following meaning: C and D are the endpoints of the cable line section that correspond to the ends of the exposed section of the planed metal installation, C' and D'; R, S, and T are the phase conductors; P are the planed metal pipeline or some other type of metal installations; and D_1, D_2, and D_3 are the distance between the planned/designed installation and the individual phase conductors and their metal sheaths/screens.

Bearing in mind its screening function concerning the inductive influence of cable lines in the further consideration instead of cable sheath will use the term cable screen.

On the basis of the previous consideration and Fig. 7.1, it is not difficult to see that the magnitude of the induced voltages and currents depends on many local factors like the number of surrounding metal installations, their shape, and dimensions, their mutual dispositions, including their spatial position toward the inducing line and the induced installation. Thus in reality, we have a very complex problem that can be defined in the following simple way: How we can take into consideration the screening effect of all, known and unknown, metal installations surrounding an HV or EHV line?

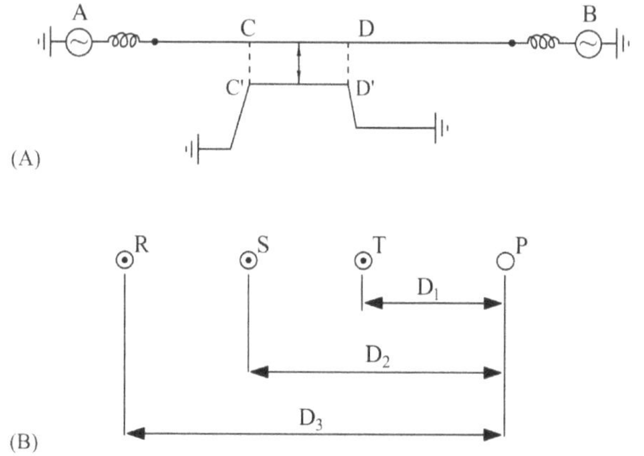

Figure 7.1 A situation typical for an inductive influence. (A) Section of the exposure. (B) Spatial position in relation to the line conductors.

For the complexity of the described problem, many projects to study possible effects from exposure to electromagnetic fields caused by electricity supply systems have been launched and some computer-based tools for the investigation on the interference have been developed. By using them, it is possible to take into account many relevant factors and parameters, including those of a lower order. However, the great calculation capabilities of modern computers cannot be of help in urban and suburban areas, where neither all input data can be determined and collected. Thus the problem considered here can be solved only by applying the methodology that compensates for the deficiency of an arbitrarily large number of relevant but unknown data. Such methodology is completely developed in Ref. [5,6] and will be presented herein.

7.2 Exposure to inductive influence along the whole cable line

7.2.1 Complete equivalent circuit of an HV or EHV cable line

On the basis of the previous consideration, each HV or EHV cable line represents a very complex electrical circuit with many conductively and inductively coupled elements. Thus the complete electrical circuit of one such power line, including one designed metal installation denoted with $(N+1)$ under the conditions of a simulated ground fault in supplied substation B, can be, by certain idealizations and simplifications, presented as shown in Fig. 7.2.

The used notation has the following meaning: U_a is the auxiliary voltage source; U_{n0}, U_{n1}, $U_{n2}, \ldots$, U_{nN} are the voltages induced in an arbitrary (nth) metal installation by the current in each of the surrounding electrical circuits; I_t is the simulated ground-fault current through one of the phase conductors of the considered line; I_1 is the current through the screen of the single-core cable carrying test current I_t; I_2, I_3 are the currents induced in the cable screens of the other two single-core cables; I_4, I_5, …, I_n, …, I_N are the currents induced in the individual surrounding metal installations; Z_0 is the self-impedance of the applied phase conductor of the considered cable line; Z_{scR}, Z_{scS}, Z_{scT} are the self-impedances of the cable screens ($Z_{scR}=Z_{scS}=Z_{scT}$); Z_4, Z_5, Z_6, …, Z_N are the self-impedances of the individual surrounding metal installations; $Z_{(N+1)}$ is the self-impedance of the designed metal installation; and N is the total number of surrounding metal installations enlarged by the number of the considered cable line screens (This number, observed theoretically, can be arbitrarily large).

As can be seen, the presented equivalent circuit is composed of the self and mutual impedances of all surrounding metal installations, all cable line screens, and phase conductor carrying current I_t, as well as all induced voltages. The self-impedance of an arbitrary, nth, surrounding metal installation is, according to, for example [20], determined by

$$Z'_n = R'_n + \omega_t\frac{\mu_0}{8} + j\omega_t\frac{\mu_0}{2\pi}\ln\frac{\delta_t}{r_n}, \quad \Omega/\text{km}, \tag{7.1}$$

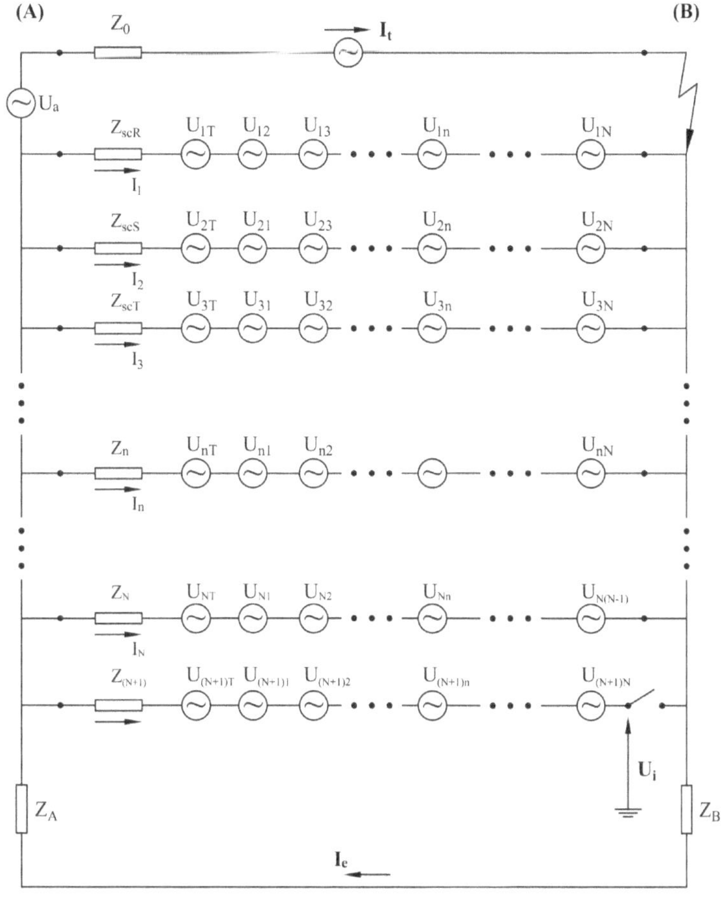

Figure 7.2 Complete equivalent circuit of a tested high-voltage cable line.

whereas mutual impedance between two arbitrary, *n*th and *m*th, surrounding metal installations is determined by

$$Z'_{nm} = \omega_t \frac{\mu_0}{8} + j\omega_t \frac{\mu_0}{2\pi} \ln \frac{\delta_t}{d_{nm}}, \quad \Omega/\text{km}; \quad m \neq n \tag{7.2}$$

where R'_n is the longitudinal resistance of an arbitrary, *n*th, surrounding metal installation, Ω/km; r_n is the mean radius of an arbitrary, *n*th, surrounding metal installation (m); d_{nm} is the distance between two arbitrary, *n*th and *m*th, surrounding metal installations (m); ω_t is the angular test frequency: $2\pi f_t$; μ_0 is the magnetic permeability of vacuum: $4\pi \cdot 10^{-7}$ Vs/Am; and δ_t is the equivalent earth penetration depth (m).

The equivalent earth penetration depth is determined by

$$\delta_t = 658\sqrt{\frac{\rho}{f_t}}, \tag{7.3}$$

where ρ is the equivalent soil resistivity along and around the cable line (Ωm), and f_t is the test circuit frequency.

Current in an arbitrary surrounding installation, I_n, induces in an also arbitrary (mth) surrounding installation a voltage, U_{mn}, which is determined by

$$U_{mn} = -Z'_{nm} I_n, \quad \text{V/km}, \quad m \neq n \tag{7.4}$$

where Z'_{nm} is the mutual impedance between the mth and nth surrounding metal installation.

Also, for further consideration, it is necessary to say that the current directions shown in Fig. 7.2 are taken arbitrarily.

By using the equivalent circuit shown in Fig. 7.2, one can write for each induced current from I_1 to I_N by the one equation and obtain a system of N independent equations. Then, based on the known values of the anticipated actual ground-fault current I_F (available on the basis of a separate study), values of all self- and mutual impedances, and this system of equations it is possible to determine each of the induced currents in the conditions of an actual ground fault. However, it is sufficient that only one of the self- or mutual impedances in the presented complete equivalent circuit is unknown and none of the induced currents can be determined and the whole problem seems unsolvable. Unfortunately, because of the previously mentioned practical limitations and impossibilities, the numerous relevant data necessary for determining the self- and mutual impedances of surrounding metal installations, even the total number of the surrounding metal installations, are uncertain or completely unknown in many practical cases.

On the basis of the equivalent circuit in Fig. 7.2, it can be seen that each induced current in the surrounding conductors (cable line screens and surrounding metal installations) appears as a consequence of the inductive influence of the test current through the line phase conductor I_t and the currents induced in all other surrounding conductors. Thus each of the presented currents cumulatively involves the inductive influences of all others, known and unknown, currents presented in Fig. 7.2. According to this, the value of each of the induced currents depends on the constructive characteristics and spatial position of each constitutive element of the spontaneously formed electrical circuit, including even those in which the induced currents are very small. These facts are, certainly, of prime importance for solving the problem of the deficiency of relevant data but are not sufficient for its solving. They only indicate in which direction the further investigations should be orientated showing the quantities that cumulatively encompass the influences of all relevant factors and parameters, including those whose significance is negligibly small.

The presented complete equivalent circuit is not without idealizations and simplifications in relation to the actual physical mode, although incorporates all, known and unknown, surrounding metal installations. The fact that the majority of the surrounding metal installations are grounded at certain places, usually at all buildings arranged along the street(s) is disregarded. Also, the fact that the metal sheaths in the case of cable lines with applied cross-bonding are grounded at each of the cross-bonding boxes has been disregarded. However, these facts are not at all relevant for the considered problem, because we analyze only the currents that appear as a consequence of the inductive couplings. Since these currents circulate only in the axial direction through the conductors, it is not relevant whether these installations are insulated toward the earth or in direct and continuous contact with the earth along their entire length. Also, because of the same reason, it is not relevant whether the equivalent conductor substituting all these installations has some insulation toward the earth or not. For a better understanding of the reasons for these neglects, it is useful to remind of a similar and in the contemporary power delivery practice well-known example. That is the case of determining the reduction factor of overhead lines. Namely, when we solve this problem, we also do not take into account that its ground wire(s) is grounded at each line tower, for example [20].

Also, as is well known HV distribution substations are mainly located in urban or suburban areas occupied by many underground metal installations, acting as perfect grounding electrodes [23]. Thus the quantitative relationships between the impedances shown in the equivalent circuit of Fig. 7.2 are such that grounding impedances Z_A and Z_B can be disregarded ($Z_A \approx 0$ and $Z_B \approx 0$) in further analysis. Based on these approximations, each induced current is a consequence solely of the mutual inductive coupling with all other surrounding electrical circuits formed by all screens of the considered line and all surrounding metal installations. However, the presented methodology is developed based on the measured currents I_t and I_1. According to that it takes into account the fact that the real values of Z_A and Z_B are somewhat larger than zero. Because of that currents I_t and I_1 contain components appearing only as a consequence of potentials $Z_A I_e$ and $Z_B I_e$. However, these components of currents I_t and I_1 are negligibly small in practical conditions because of the following reasons:

- current I_e is only a few percents of I_t because of strong inductive coupling between the phase conductor with current I_t on the one side and the cable line screens, including all surrounding metal installations on the other side [6], and
- an HV or EHV cable line is so long that almost each HV distribution substation and its feeding cable line satisfy the following inequality:

$$Z_A \approx Z_B < Z_1' L, \tag{7.5}$$

where L is the total cable line length (km), and Z_1' is the self-impedance of cable screen per unit length (Ω/km).

Besides, by the circuit in Fig. 7.2 is disregarded the fact that the relevant constructive parameters of some installations are not quite uniformly distributed along the length of these installations, as well as that all sections of one installation usually are not completely in parallel in relation to the considered cable line and other of the surrounding metal installations. However, these neglects are also not at all relevant for the development of the here-presented methodology. The development of the methodology of compensation of deficiency of all relevant but unknown data is based on only the values of currents I_t and I_1 in Fig. 7.2. All the other relevant data concerning the surrounding metal installations are included through only these two currents as shown in Chapter 3, Actual reduction factor of power distribution lines.

Also, the balanced currents passing through the phase conductors in the conditions of an actual ground fault are disregarded in this equivalent circuit. However, the influence of the ground-fault current components passing only through phase conductors on the surrounding metal installations is, as is well known, practically negligible. Something similar can be said for the capacitive currents in the considered cable line during the necessary test measurements. They are disregarded because voltage U_a is very small (usually between 100 and 300 V) in comparison with the nominal voltages of HV and EHV cable lines.

Finally, the soil around and along an HV cable line is normally heterogeneous, but when we calculate impedances Z_n and Z_{nm}, we assume that it is homogeneous and that its resistivity is equal to its equivalent soil resistivity, ρ. Fortunately, these impedances are, according to Eqs. (7.1) and (7.2), only slightly dependent on the equivalent soil resistivity. Thus this idealization of the actual environment, unavoidable because of the well-known reasons, can be applied without noticeable sacrificing of the accuracy of the analytical part of the here-presented methodology.

7.2.2 Equivalent cable line screen

In urban conditions, metal installations surrounding an HV or EHV cable line are mainly situated under the surface of the ground and according to that are not accessible for any measurements. Thus many of the currents presented in Fig. 7.2 cannot be determined by direct measurements. Fortunately, this is not the case with the currents passing through cable line conductors. This fact enables us to obtain data about the cumulative inductive influence of all surrounding metal installations, including cable line screens on the value of the current circulating exclusively through the earth, I_e. This becomes possible when we by measurements obtain two of the currents shown in Fig. 7.2, the test current through the phase conductor of the used single-core cable, I_t, and the current through the screen belonging to this single-core cable, I_1. These currents can be determined by using the measurement circuit that is principally shown in Fig. 7.3.

The used notation has the following meaning: *A* is the ampere-meter.

With the aim of simplifying the complex equivalent circuit in Fig. 7.2, let us introduce an equivalent conductor substituting all surrounding metal installations, including screens of the two, for the measurements, unused single-core cables. Then, the complete equivalent circuit of the considered cable line, by using approximations $Z_A \approx 0$ and $Z_B \approx 0$ can be simplified and presented as shown in Fig. 7.4.

The used notation has the following meaning: U_{10} (U_{1eq}) is the voltage that current I_t (I_i) induces in the cable screen with current I_1; U_{eq0} (U_{eq1}) is the voltages that current I_t (I_1) induces in the equivalent conductor; Z_0 is the self-impedance of the applied phase conductor of the considered cable line; Z_1 is the self-impedance of the cable screen with I_1 (in Fig. 7.2 denoted as Z_{scT}); and Z_{eq} is the self-impedance of the introduced equivalent conductor.

Based on the equivalent circuits presented in Figs. 7.2 and 7.4, it is clear that many $(N-1)$ unknown currents through the surrounding metal installations, including the cable line screens, whose total number observed theoretically can be infinitely large,

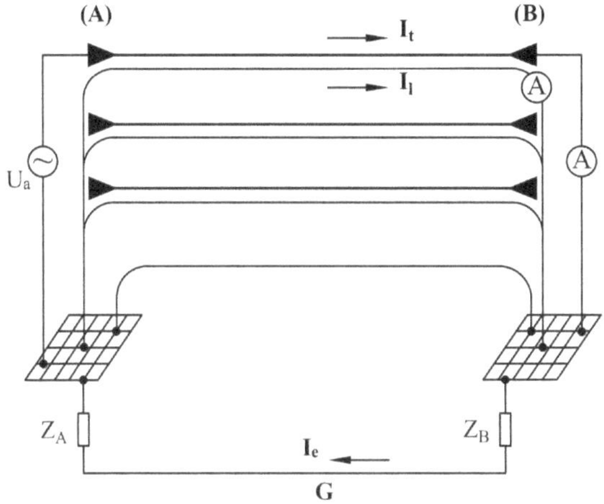

Figure 7.3 The principal measurement circuit.

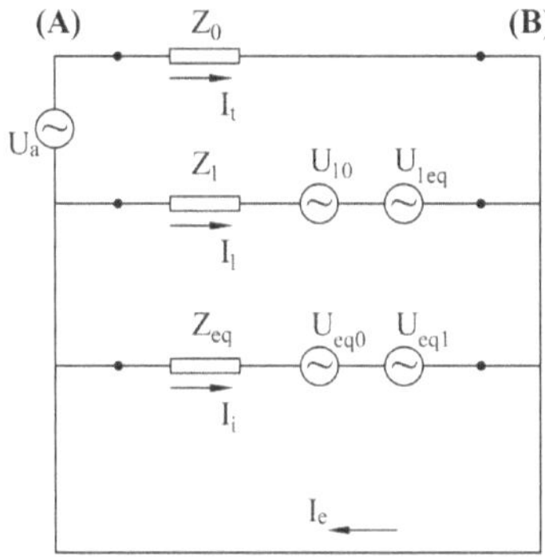

Figure 7.4 Equivalent circuit with an equivalent conductor.

are reduced to only one unknown current, I_i. Also, numerous known and unknown metal installations, as well as screens of the considered line encompassed by the fluctuating magnetic field around and along the used phase conductor, are substituted by only one equivalent conductor. In this way the numerous relevant but unknown data are reduced to the relevant data concerning only this, equivalent conductor.

The relevant parameters of the introduced equivalent conductor can be determined by using the condition that the currents I_t, I_1, and I_e (Figs. 7.2 and 7.4) have to remain unchanged by introducing this conductor. This condition can be, according to the equivalent circuit in Fig. 7.4, expressed by

$$\begin{aligned} &1.\ Z'_{10}I_t + Z'_1 I_1 + Z'_{1eq}I_i = 0 \\ &2.\ Z'_{eq0}I_t + Z'_{eq1}I_1 + Z'_{eq}I_i = 0 \end{aligned} \tag{7.6}$$

where Z'_{eq0} is the mutual impedance between the introduced equivalent conductor and the phase conductor of the cable with current I_t, per unit length (Ω/km); Z'_{1eq} is the mutual impedance between the introduced equivalent conductor and the screen of the cable with current I_t, per unit length (Ω/km).

In Eq. (7.6), impedances Z'_1 and Z'_{10} are, according to Ref. [20], determined by

$$Z'_1 = R'_{sc} + \omega_t \frac{\mu_0}{8} + j\omega_t \frac{\mu_0}{2\pi} \ln \frac{\delta_t}{r_{sc}}, \quad \Omega/\text{km}, \tag{7.7}$$

$$Z'_{10} = \omega_t \frac{\mu_0}{8} + j\omega_t \frac{\mu_0}{2\pi} \ln \frac{\delta_t}{r_{sc}}, \quad \Omega/\text{km}. \tag{7.8}$$

where R'_{sc} is the resistance of the cable line screen per unit length (Ω/km), and r_{sc} is the mean radius of the cable line screen (m).

With the aim of minimizing the number of unknown but relevant parameters of the introduced equivalent conductor, this conductor is imagined as a cylinder around the used single-core cable and along the entire line. According to that instead of the term equivalent conductor, we can use the term equivalent screen in further consideration. Here, it is also important to mention that for any other imagined shape and spatial position of this conductor, the considered problem would be unsolvable because the number of its relevant but unknown data would be too large.

On the basis of the introduced equivalent conductor imagined in such a way, the analytical expressions for impedances Z_{eq}, Z_{0eq}, and Z_{1eq} become in advance known and, according to Ref. [20], determined by

$$Z'_{eq} = R'_{eq} + \omega_t \frac{\mu_0}{8} + j\omega_t \frac{\mu_0}{2\pi} \ln \frac{\delta_t}{r_{eq}}, \quad \Omega/\text{km}, \tag{7.9}$$

$$Z'_{0eq} = Z'_{1eq} = \omega_t \frac{\mu_0}{8} + j\omega_t \frac{\mu_0}{2\pi} \ln \frac{\delta_t}{r_{eq}}, \quad \Omega/\text{km}, \tag{7.10}$$

where R'_{eq} is the resistance of the introduced equivalent screen per unit length (Ω/km), and r_{eq} is the mean radius of the imagined cylinder representing the introduced equivalent screen (m).

Also, since currents I_t and I_1 are obtained by measurements, they can be treated as known quantities, and condition defined by Eq. (7.6) can be modified into the following:

$$\frac{Z'_{0eq} Z'_{1eq} - Z'_{eq} Z'_{01}}{Z'_{eq} Z'_1 - Z'^2_{1eq}} = \frac{I_1}{I_t}. \tag{7.11}$$

Since Eq. (7.11) gives the relationship between complex quantities, it can be presented as the following system of equations:

$$\begin{aligned} &1.\ \text{Re}\left\{\left(Z'_{0eq} Z'_{1eq} - Z'_{eq} Z'_{01}\right) I_t\right\} = \text{Re}\left\{\left(Z'_{eq} Z'_1 - Z'^2_{1eq}\right) I_1\right\} \\ &2.\ \text{Im}\left\{\left(Z'_{0eq} Z'_{1eq} - Z'_{eq} Z'_{01}\right) I_t\right\} = \text{Im}\left\{\left(Z'_{eq} Z'_1 - Z'^2_{1eq}\right) I_1\right\} \end{aligned} \tag{7.12}$$

where Re and Im are the real and imaginary parts of Eq. (7.11).

In these equations the only unknown quantities, according to Eqs. (7.7)–(7.10), are parameters R'_{eq} and r_{eq} and according to this can be determined by solving this system of equations.

After determining the relevant parameters of the equivalent screen (R'_{eq} and r_{eq}), Eqs. (7.9) and (7.10) and the equivalent circuit of Fig. 7.4 can be used for obtaining impedances: Z_{eq}, Z_{0eq}, and Z_{1eq}, as well as currents: I_i and I_e. Then, bearing in mind the definition of the screening factor of line neutral (screening) conductors [19], the resulting screening factor is determined by

$$k_r = \frac{I_e}{I_t} = 1 - \frac{\left(Z'_{eq} - Z'_{1eq}\right)Z'_{01} + \left(Z'_1 - Z'_{1eq}\right)Z'_{0eq}}{Z'_1 Z'_{eq} - Z'^2_{1eq}}, \tag{7.13}$$

or, according to Eqs. (7.7) and (7.8), in a somewhat more compact form:

$$k_r = 1 - \frac{R'_{eq}Z'_{01} + \left(Z'_1 - Z'_{1eq}\right)Z'_{0eq}}{Z'_1 Z'_{eq} - Z'^2_{1eq}}. \tag{7.14}$$

Now, we will make one more simplification by introducing an equivalent screen that encompasses the previously introduced equivalent screen and cable line screen with current I_1. By this substitution the considered cable line, including all surrounding metal installations, is modified into the very simple line model consisting of only one single-core cable whose cross section is shown in Fig. 7.5.

R'_{es} is the resistance of the newly introduced equivalent screen per unit length (Ω/km), and r_{es} is the mean radius of the imagined cylinder representing the newly introduced equivalent screen (m).

Then, on the basis of the analytical expression for the reduction/screening factor of one single-core cable line shown in Fig. 7.5, we obtain the relationship between the resulting screening factor k_r and the parameters of the finally introduced equivalent screen, R'_{es} and r_{es}. This relationship is, according to Eq. (2.10) in Chapter 2, determined by the following system of equations:

$$\begin{aligned} &1.\ \mathrm{Re}\left\{k_r\left(R'_{es} + \omega\frac{\mu_0}{8} + j\frac{\omega\mu_0}{2\pi}\ln\frac{\delta}{r_{es}}\right)\right\} = R'_{es} \\ &2.\ \mathrm{Im}\left\{k_r\left(R'_{es} + \omega\frac{\mu_0}{8} + j\frac{\omega\mu_0}{2\pi}\ln\frac{\delta}{r_{es}}\right)\right\} = 0. \end{aligned} \tag{7.15}$$

where Re and Im are the real and imaginary parts of Eq. (2.10) in Chapter 2.

However, this system of equations has an unlimited number of solutions for r_{es} and corresponding R'_{es} and vice versa. By adopting that the equivalent screen has the same physical appearance as the actual cable screen, this fact gives us the possibility to take into account the inductive influence of surrounding metal installations,

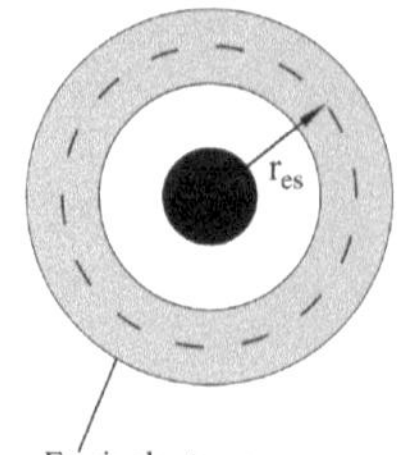

Figure 7.5 Cross section of the finally simplified line model.

including cable line sheaths through only one cable sheath parameter without changing its physical appearance and spatial position. In other words, by adopting $r_{es} = r_{sc}$, the physical appearance and spatial positions of actual cable line conductors remain unchanged (Fig. 7.1) and only parameter that should be determined is the equivalent longitudinal resistance of the cable sheath with current I_1, $R'_{es}(r_{es} = r_{sc})$. It is, according to Eq. (7.15), determined by

$$\mathrm{Im}\left\{k_r\left(R'_{es}(r_{es} = r_{sc}) + \omega\frac{\mu_0}{8} + j\frac{\omega\mu_0}{2\pi}\ln\frac{\delta}{r_{sc}}\right)\right\} = 0. \tag{7.16}$$

After that, also on the basis of expression for the reduction factor of the cable line consisting of only one single-core cable Eq. (7.10), we obtain one more expression for the resulting screening factor that is

$$k_r(r_{es} = r_{sc}) = \frac{R'_{es}(r_{es} = r_{sc})}{R'_{es}(r_{es} = r_{sc}) + \omega(\mu_0/8) + j(\omega\mu_0/8)\ln(\delta/r_{sc})} \tag{7.17}$$

As can be seen based on Eq. (7.17), the total influence of all surrounding metal installations, including cable line screens is taken into consideration through the value of only one parameter of the cable screen with current I_1, $R'_{es}(r_{es} = r_{sc})$. Certainly, here it should be mentioned that this is just an apparent parameter (expressed by a complex number) but can be useful for further consideration of the inductive influence of HV and EHV lines. For instance, at determining the inductive influence of an HV and EHV cable lines during normal operating conditions we can, based on $R'_{es}(r_{es} = r_{sc})$ and the principle of superposition use so far known and relatively simple calculation procedure [19].

Here it is also interesting to note that the presented analytical procedure can be interpreted in one more way, somewhat nearer to the actual physical phenomenon. Namely, in reality, each of the cable line screens and each of the surrounding metal installations with the earth as a common return path act as a short-circuited secondary winding with only one turn of the spontaneously formed, very large and specific transformer. By using this analogy the previously defined and used condition for determining the equivalent cable screen substituting all cable line screens and all surrounding metal installations (R'_{es} and r_{es}) can be defined in one more way. The resulting alternating magnetic flux through the primary winding with one turn (the surface closed by the phase conductor with current I_t and the earth as a common return path) should remain unchanged after introducing the equivalent screen (equivalent secondary winding consisting also of only one turn). For a certain HV or EHV cable line and certain current I_t, this flux depends only on the imagined spatial position of the equivalent cable screen (equivalent secondary turn), r_{es}, in relation to the phase conductor (primary turn) and the current through this screen, I_i. Since this current depends on R'_{es}, the required condition can be satisfied with an arbitrarily large number of different combinations of r_{es} and corresponding R'_{es}. The observations and conclusions that emanate from the analogy with this specific transformer are considered in more detail in Chapter 3, Actual reduction factor of power distribution lines. This consideration, certainly, represents one more proof for the validity of the here-presented methodology.

7.2.3 Cable line with no applied cross-bonding

7.2.3.1 Ground-fault conditions

When the resulting screening factor for the nearest phase conductor (Fig. 7.1) can be determined, the voltage induced at the end of the planned metal installation (Fig. 7.2) can be, according to [19], determined by

$$U_i(I_F) = k_r L Z'_{T(N+1)} I_F \quad \text{(V)}, \tag{7.18}$$

where I_F is the maximal anticipated ground-fault current in supplied substation B (A); $Z'_{(N+1)T}$ is the mutual impedance between the nearest phase conductor and the designed metal installation (Ω/km); and L is the length of exposure (the whole length of the induced line that is approximately equal to the length of the inducing line or longer).

Thus when current I_F is known, for the determination of the induced voltage at the end of the planned installation (Fig. 7.2), it is sufficient to know the spatial position of the designed installation in relation to the faulty phase conductor. Then, for the known constructive characteristics of the designed installation, the current that will be induced in this installation is determined by

$$I_i(I_F) = \frac{k_r L Z'_{T(N+1)} I_F}{Z_{(N+1)}} \quad \text{(A)} \tag{7.19}$$

As can be seen by using expression (7.17), solving the problem of the inductive influence of cable line during a ground fault is reduced to the simplest case where the inducing line consists of only one single-core cable.

7.2.3.2 Normal operation conditions

The complete equivalent circuit of an HV or EHV cable line under normal operating conditions is presented in Fig. 7.6.

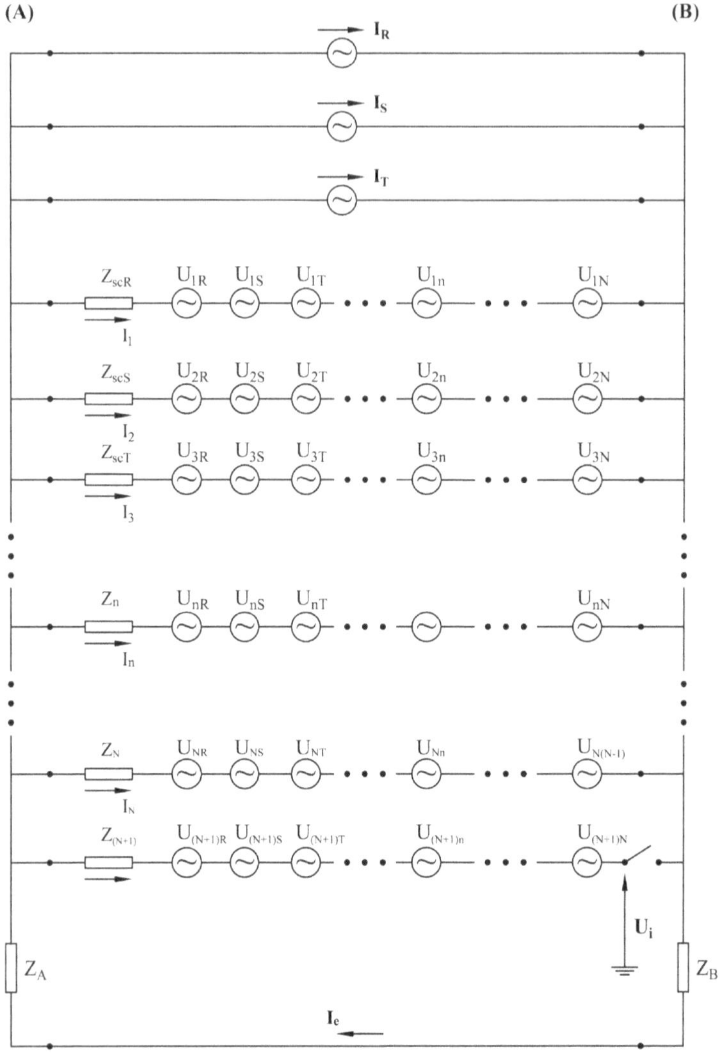

Figure 7.6 Complete equivalent circuit of a cable line under normal operating conditions.

According to this equivalent circuit, each of the phase conductors acts as an individual inducing line so that by applying the principle of superposition we obtain

$$U_i(I_R, I_S, I_T) = L\left(k_R Z'_{(N+1)R} I_R + k_S Z'_{(N+1)S} I_S + k_T Z'_{(N+1)T} I_T\right) \quad \text{(V)} \tag{7.20}$$

$$I_i(I_R, I_S, I_T) = \frac{L\left(k_R Z'_{(N+1)R} I_R + k_S Z'_{(N+1)S} I_S + k_T Z'_{(N+1)T} I_T\right)}{Z_{(N+1)}} \quad \text{(A)} \tag{7.21}$$

where k_R, k_S, k_T is the resulting screening factor for each phase conductor seen separately, and I_R, I_S, I_T are the maximal load currents in the phase conductors.

In many practical cases the single-core cables are mutually closed (touch each other), and resulting screening factors for different phase conductors k_R, k_S, and k_T have practically the same value, or: $k_R \approx k_S \approx k_T$.

On the basis of expressions (7.20) and (7.21), it is clear that they can be used for any practically possible loads (I_R, I_S, I_T), including those that are extremely unbalanced, that is, that are the most critical from the standpoint of the magnitude of voltages and currents induced in surrounding metal installations.

In this case by using expression (7.17), solving the problem of the inductive influences of cable line under normal operating conditions is reduced to the simplest case where only phase conductors and screens of the cable line exist, that is, where the surrounding metal installations do not exist.

7.2.4 Cable line with applied cross-bonding

7.2.4.1 Ground-fault conditions

The length of HV and EHV cable lines in a distribution network is so large that the application of the cross-bonding technique is required for each of them [19]. When the cross-bonding is applied, the line route is divided into major sections, each of which is divided into three subsections. In each of these subsections, the three screens connected in series are associated with conductors of different phases, and consequently, the voltages of the screens have the same magnitude but with a phase shift of 120 degrees. The overall effect achieved by cross-bonding is that the resultant voltage induced during normal operating conditions in each of the cable screens is equal to zero so that through the cable screens there are no the so-called circulating currents.

Due to the described mutual exchanges of the cable screens (cross-bonding technique), the considered inductive influences during a simulated ground fault seem this time significantly more complicated (Fig. 7.2) and the relatively simple analytical expressions for the mutual impedances between a phase conductor and the screens of the cable line become more complicated. Also, since the cable screens are grounded at each cross-bonding box, current I_1 obtained by test measurement would not be a consequence of only inductive couplings. As such this current would not be used for determining the parameters of the introduced equivalent screen and the application of the previously described methodological version (Section 7.2.3). Because of that, for obtaining the necessary test circuit, this time we will use another favorable circumstance concerning HV and EHV cable lines. The phase conductors of these lines, as is well known, are coated by an insulating layer and their mutual spatial position is unchangeable along the entire line. Therefore for obtaining current I_1, this time we will use one of two previously unused phase conductors as shown in Fig. 7.7.

The used notation has the following meaning: D is the cable screen disconnection from the grounding electrode of substation B, and I_2, I_3, and I_4 are the currents induced in the cable screens.

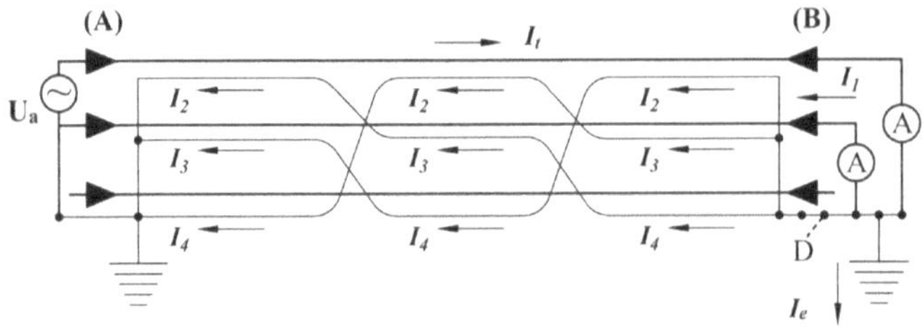

Figure 7.7 Measurement circuit for cable lines with cross-bonding.

A complex electrical circuit that is spontaneously formed along and around the test circuit we will this time also simplify by introducing an equivalent conductor that substitutes all three cable line screens and all the surrounding metal installations. In this way, in reality, a complex electrical circuit is substituted by a relatively simple equivalent circuit that is the same as the circuit shown in Fig. 7.4 only certain notations have somewhat changed meanings. According to the measuring circuit in Fig. 7.7, the used notation has the following meaning: U_{10} (U_{1eq}) is the voltage that current I_t (I_i) induces in the additional phase conductor with current I_1; U_{eq0} (U_{eq1}) are the voltages that current I_t (I_1) induces in the equivalent conductor; Z_0 is the self-impedance of the phase conductor with current I_t; Z_1 is the self-impedance of the additionally used phase conductor with current I_1; and Z_{eq} is the self-impedance of the introduced equivalent conductor.

Using the equivalent circuit in Fig. 7.4, the relevant parameters of the equivalent conductor can be also determined using the conditions defined by the system of Eq. (7.6). It is only necessary to take care that the impedances Z_1' and Z_{01}' are, according to the introduced changes, here determined by the following expressions:

$$Z_1' = Z_{ph}' = R_{ph}' + \omega_t \frac{\mu_0}{8} + j\omega_t \frac{\mu_0}{2\pi}\left(\frac{\mu_r}{4} + \ln\frac{\delta}{r_{ph}}\right), \quad \Omega/\text{km}, \tag{7.22}$$

$$Z_{01}' = Z_{1eq}' = \omega_t \frac{\mu_0}{8} + j\omega_t \frac{\mu_0}{2\pi} \ln\frac{\delta_t}{D_{ph}}, \quad \Omega/\text{km}. \tag{7.23}$$

where R_{ph}' is the resistance of the phase conductor per unit length (Ω/km), and D_{ph} is the distance between the phase conductors applied for the necessary measurements (m).

Then, by using the same procedure as the one previously described, we obtain that the resulting screening factor is also determined by Eq. (7.14). Since the phase conductor used for obtaining current I_1, during an actual ground fault does not participate in ground-fault current returning, this conductor cannot be taken into account in determining the equivalent screen. Thus this time, we have to introduce in Eq. (7.14) the corresponding values concerning this conductor: $Z_1 \to \infty$, $R_1 \to \infty$,

$Z_{01}=0$, and $Z_{1mi}=0$ with the aim of its elimination. According to that, instead of Eq. (7.14), we obtain

$$k_r=\frac{Z'_{eq}-Z'_{0eq}}{Z'_{eq}}, \tag{7.24}$$

Then, based on Eqs. (7.15) and (7.16), we can determine the equivalent longitudinal resistance of the actual cable screen substituting the inductive influence of all other cable screens and all surrounding metal installations, $R'(r_{eq}=r_{sc})$. In this way, we are replacing during a ground fault an actual cable line and all surrounding metal installations with one of its single-core cables, but screen of which is with an equivalent longitudinal resistance. The value of this resistance of the screen of the single-core cable with current I_t is such that the inductive influences of all three screens of the cable line and all surrounding metal installations are encompassed through it. After that, expression (7.24) can be presented in the following, more direct form:

$$k_r(r_{eq}=r_{sc})=\frac{R'_{eq}(r_{eq}=r_{sc})}{R'_{eq}(r_{eq}=r_{sc})+\omega_t(\mu_0/8)+j\omega_t(\mu_0/2\pi)\ln(\delta_t/r_{sc})}. \tag{7.25}$$

Based on the known resulting screening factor, the voltages and currents induced during ground fault are obtained by using Eqs. (7.18) and (7.19). Certainly, the critical phase conductor, from the standpoint of the magnitude of inductive influence is the one that is nearest to the designed metal installation.

As well as in the case considered in Section 7.2.3.1, in this case by using expression (7.25) solving the problem of the inductive influence of cable line during a ground fault is reduced to the simplest case where only one single-core cable exists, Fig. 7.5.

Finally, the flowchart describing the whole here-presented calculation procedure for the determination of the induced voltage and current is shown in Fig. 7.8.

7.2.4.2 *Normal operation conditions*

Because of the applied cross-bonding technique, during normal operating conditions, the currents through the cable line screens undo mutually so that the sum of currents I_2, I_3, and I_4 (Fig. 7.7) is practically equal to zero, $I_2+I_3+I_4\approx 0$. Thus the screening effect exists only as a consequence of the presence of surrounding metal installations, whereas cable line screens do not participate in creating this effect. Because of that during the test measurements, it is necessary to disconnect the screens of all three cables from one of the grounding electrodes at the line ends (*D* in Fig. 7.7).

By omitting the cable screens the complete electrical circuit obtained during the measurements can be imagined on the basis of Figs. 7.6 and 7.7. Then, by introducing an equivalent cable screen substituting all surrounding metal installations, this

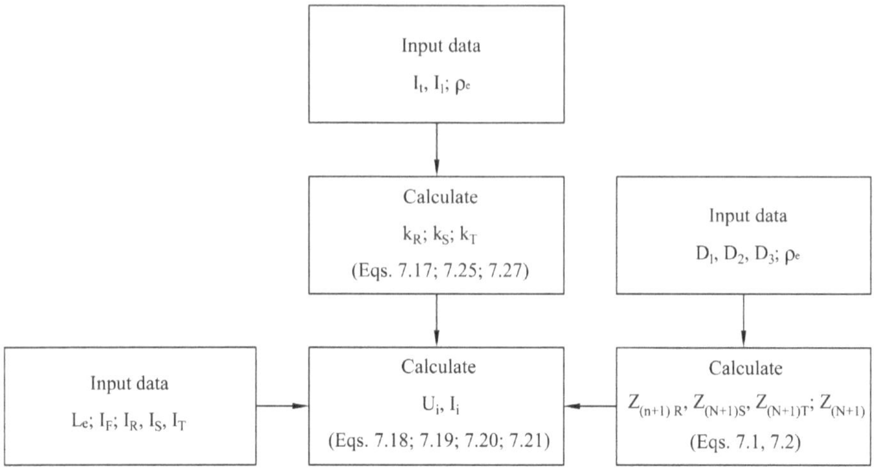

Figure 7.8 Flow diagram of necessary calculations.

complex circuit can also be replaced with the simple equivalent circuit shown in Fig. 7.4. At this the used notations retain the same meanings, only the equivalent screen, in this case, does not encompass the screens of the considered cable line.

Based on the equivalent circuit in Fig. 7.4, the relevant parameters of the equivalent cable screen, in this case, can be also determined using the condition defined by Eq. (7.6) and the same procedure as the one described in Section 7.2.2. In this way, we obtain that the resulting screening factor is also determined by Eq. (7.14).

Since the phase conductor used for current I_1, in reality, does not represent one of the neutral conductors, for its elimination we have to introduce in Eq. (7.14) the following values concerning this conductor: $Z_1 \to \infty$, $R_1 \to \infty$, $Z_{01} = 0$, and $Z_{1mi} = 0$. Then, we obtain that the resulting screening factor of all surrounding metal installations, k_{mi} is determined by

$$k_{mi} = \frac{Z'_{mi} - Z'_{0mi}}{Z'_{mi}}. \tag{7.26}$$

or, in a more direct form:

$$k_{mi} = \frac{R'_{mi}}{R'_{mi} + \omega_t(\mu_0/8) + j(\omega_t\mu_0/2\pi)\ln(\delta_t/r_{mi})}. \tag{7.27}$$

where R'_{mi} is the longitudinal resistance of the equivalent screen substituting all surrounding metal installations (Ω/km), and r_{mi} is the mean radius of the equivalent screen substituting all surrounding metal installations (m).

The longitudinal resistance of the equivalent sheath R'_{mi} can be determined using expression (7.15) for an arbitrarily adopted value of mean radius r_{mi}. As well as in the previously considered case, the resulting screening factor, r_{mi}, should be determined for each of the phase conductors, seen separately. Then, the induced voltage and current are determined by using Eqs. (7.20) and (7.21).

Since screening factor k_{mi} is significantly larger than k_r ($k_{mi} >> k_r$), the inductive influence of an HV or EHV cable line under normal operating conditions is significantly intensified when cross-bonding is applied. Thus it is not difficult to see that a significant reduction, or practical elimination, of the harmful circulating currents in cable line screens, achieved by cross-bonding, unavoidably intensifies the harmful currents circulating through the surrounding metal installations (Fig. 7.6). However, energy losses in surrounding metal installations are certainly less harmful than those in cable screens that cause additional cable heating and reducing their ampacity.

As well as in the case considered in Section 7.2.3.2, in this case by using expression (7.27), solving the problem of the inductive influence of cable line during normal operating conditions is reduced to the simplest case where only the cable line and its conductors exist, that is, where the surrounding metal installations are substituted with the equivalent cable line screens.

Since the surrounding metal installations are not uniformly distributed along the inducting line, the resulting screening factors determined by the presented expressions are quite correct only in cases when the induced line is exposed along the entire length of the inducting line. In all other cases, they can be used only for one approximate prediction of the voltage and current that will be induced in the designed installation.

7.3 Exposure to inductive influence along any section of a cable line

7.3.1 *Critical ground-fault position*

For determination the critical (maximal) voltage that could be induced in the planned installation, it is necessary to know the anticipated maximal ground-fault current through the HV or EHV cable line, I_F, and its distribution at any place along the line section from D to B (Fig. 7.1). The critical fault position, in this case, is not obvious as in the case considered in Section 7.3.2. Although the largest value of a ground-fault current along the exposure section is obviously for a fault at point D, this place is not critical from the standpoint of the inductive influence of a cable line.

Ground-fault current for an arbitrary fault place along a power lane is determined by

$$I_F(l_{AF}) = \frac{3U_{ph}}{Z_A^+ + Z_A^- + Z_A^0 + l_{AF}\left(2z_l^+ + z_l^0\right)} \tag{7.28}$$

where U_{ph} is the phase voltage at supply substation A; Z^+_A, Z^-_A, and Z^0_A are the positive, negative, and zero-sequence impedance of the power system at substation

A; z^{+}_{l}, and z^{0}_{l} are the positive and zero-sequence impedance of the cable line, per unit length (Ω/km); and l_{AF} is the cable line length from supply substation A to fault place F (km).

On the other side, the ground-fault current relative distribution for fault at any place along an HV cable line is, according to [24], determined by the following formula:

$$r_a(l_{AF}) = \frac{I_e(l_{AF})}{I_F(l_{AF})} = \frac{l_{AF}}{L} r_a(L) \tag{7.29}$$

where r_a is the actual reduction factor of the considered cable line.

Bearing in mind that for a certain line, its length and actual reduction factor, $r_a(L)$, represent constant parameters, the reduction factor for different fault places along the cable line is, according to Eq. (7.29), a linear function of the line length to the fault place, l_{AF}. The actual reduction factor of an HV or EHV cable line involves the inductive influence of all surrounding metal installations, and its value depends on all of them as shown in Chapter 3, Actual reduction factor of power distribution lines. However, the manner of the ground-fault current relative distribution for different fault places along an HV or EHV cable line does not depend on the total number of the cable line sheaths and surrounding metal installations, N (Fig. 7.2).

Since the screening factor of the line neutral (screening) conductors and reduction factor of the line are in fact the same coefficient [19,20], on the basis of Eq. (7.29) we can also write

$$k_r(l_{AF}) = \frac{l_{AF}}{L} k_r(L) \tag{7.30}$$

Based on Eqs. (7.17), (7.25), (7.27), (7.28), and (7.30), it can be seen that the critical ground fault position, in this case, is the same as in the case considered in Section 7.2, the supplied substation. In accordance with this, the previously derived analytical expressions of the resulting screening factor are relevant in this case, as well. However, these expressions can be considered correct ones for any section of exposure only under the assumption that all surrounding metal installations are uniformly distributed along the whole cable line length. Since in practice this, usually, is not the case, they can be used only for an approximate estimation of voltages and currents induced along the exposure section of an arbitrary length, L_e (from C' to D' in Fig. 7.1).

7.3.2 On-site measurements and resulting screening factor

The considered problem is specific because the resulting screening factor determined previously on the basis of an equivalent cable screen(s) of the whole inducing line, in this case, is not quite correct. As it is well known the surrounding metal installations are not quite uniformly distributed along an HV or EHV cable line and even their total number does not have to be the same along the entire length of such a line. Thus the previously derived analytical expressions for the resulting screening factor can be considered the correct one only in the case where an induced installation is exposed along the whole HV or EHV cable line. In the cases where the induced installation is exposed only along one section of the inducing line, its value depends on the spatial position of the planned installation in relation to the other surrounding metal installations. For example, if in one case the designed installation is foreseen on a side of the cable line on which all other installations are laid, the resulting screening effect will be larger than in the case if this installation is on the other side of the inducing line in relation to all other metal installations. Thus for more correct determining the induced voltages and currents, in this case, we have to determine the resulting screening factor that relates only to the considered installation and its section of exposure.

Data regarding the magnitude of the inductive influence of an HV or EHV cable line are usually necessary at the design stage of a future nearby metal installation. For a certain inducing line and a certain ground-fault current, this influence is dependent, inter alia, on the constructive characteristics of many surrounding metal installations and their spatial positions. Since in urban and suburban conditions all necessary data cannot be collected, the possibility for solving the problem was seen in the fact that the total voltage induced in each of the surrounding installations contains the cumulative inductive influence of all other known and unknown surrounding metal installations (Fig. 7.2). Thus the problem of deficiency of many relevant data can be overcome only by performing the corresponding on-the-site measurements and obtaining this cumulative effect. However, the induced installation does not yet exist in the design stage. Because of that for obtaining the cumulative inductive influence, it is necessary to perform measurements by using one provisory cable line [5]. It can be done by laying one insulated cable on the surface of the soil exactly along the foreseen section of the exposure, from C' to D', Fig. 7.1. For this purpose, one can use a flexible (with no metal sheath) LV cable, suitable for application in different urban conditions and circumstances. This provisory cable line should be grounded at one end with the aim of forming, together with the earth as the current return path, the open electrical circuit schematically presented in Fig. 7.9.

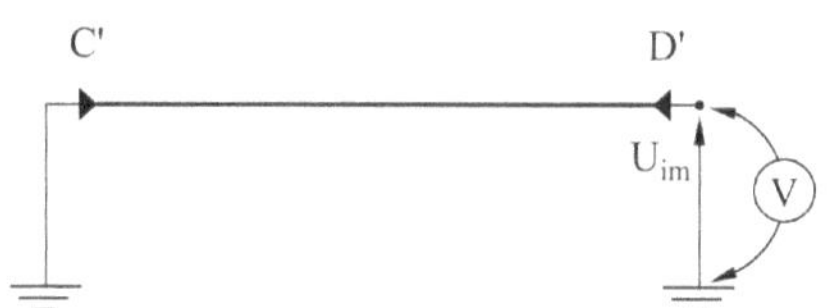

Figure 7.9 The provisory cable line for the necessary measurements.

The used notation has the following meaning: U_{im} is the total voltage (EMF) induced in the test circuit (V), V is the voltmeter, and C' (D') are endpoints of the exposed section.

From the equivalent circuit in Fig. 7.2, the total voltage induced in the provisory cable line, denoted as ($N+1$), is equal to the potential at the end of this line toward the earth, Fig. 7.9. This potential (voltage) is, according to Fig. 7.2, determined by

$$U_{im} = L_{C'D'} Z'_{(N+1)T} I_t + L_{C'D'} \sum_{n=1}^{n=N} Z'_{(N+1)n} I_n \tag{7.31}$$

where I_t is the test current obtained through phase conductor T by simulating a ground fault in supplied substation B (Fig. 7.1); $Z_{(N+1)T}$ is the mutual impedance between the planned metal installation and the nearest phase conductor, T; and $L_{C'D'}$ is the length of the exposure.

Since impedance $Z'_{(N+1)T}$ can be obtained by calculation when the spatial position of the designed installation in relation to phase conductor T is known (D_1 in Fig. 7.1) and since quantities U_{im} and I_t are obtained by measurements, all these quantities can be considered in advance known. Then, according to Eq. (7.31), the total voltage induced due to the surrounding metal installations, including the neutral conductors of the considered line is

$$\sum_{n=1}^{n=N} Z_{(N+1)n} I_n = U_{im} - Z_{(N+1)T} I_t \tag{7.32}$$

The magnitude of the screening effects is expressed as the amount of reduction of the value of the electrical quantity which is considered representative of the induction in the inducted line due to the screening current(s) [19]. Thus according to Eq. (7.31), the resulting screening factor is

$$k_r = 1 + \frac{\sum_{n=1}^{n=N} Z_{(N+1)n} I_n}{Z_{(N+1)T} I_t}. \tag{7.33}$$

Then, based on Eq. (7.32) expression (7.33) takes the following simple form:

$$k_r = \frac{U_{im}}{Z_{(N+1)T} I_t}. \tag{7.34}$$

Since metal installations in urban areas are mainly under the soil surface, the resulting screening factor obtained by using Eq. (7.34) is slightly smaller than the actual one in the case of an overhead line and slightly greater than the actual one in the case of a cable line.

When the value of the resulting screening factor is determined by Eq. (7.34), the maximal voltage and current that will be induced in the designed metal installation during a ground fault can be predicted:

$$U_i = k_r Z_{(N+1)T} I_F, \tag{7.35}$$

$$I_i = \frac{k_r Z_{(N+1)T} I_F}{Z_{(N+1)} + Z_{C'} + Z_{D'}} \tag{7.36}$$

where $Z_{C'}$ ($Z_{D'}$) is the grounding impedance at the end of the exposure, C' (D') in Fig. 7.1. I_F is the anticipated maximal ground-fault current in substation B.

Once the value of the resulting local (relevant only for the considered exposure section) screening factor is determined, to determine the voltages that will be induced in the designed installation during a ground fault, according to Eq. (7.35), it is necessary to calculate impedance $Z_{(N+1)T}$ and to know current I_F. Then, to determine the current that will be induced in this installation during a ground fault, it is necessary, in accordance with Eq. (7.36), to determine the impedances $Z_{(N+1)}$, $Z_{C'}$, and $Z_{D'}$, as well. Impedance $Z_{(N+1)}$ can be determined based on constructive parameters of the designed installation and Eq. (7.1), whereas impedances $Z_{C'}$ and $Z_{D'}$ can be determined on the basis of the field measurements. According to this, the influences of the earth as a conductive media and the influences of all surrounding metal installations, including all neutral conductors of the considered HV or EHV line are cumulatively involved through the local resulting screening factor. Thus according to the presented procedure, the determination of the inductive influence in any concrete case is reduced to the simplest case where only one induced and one screening conductor exists. Also, the relevant parameters of the equivalent screening conductor substituting surrounding metal installation and neutral conductors of the inducing line are not necessary.

Certainly, an analogous procedure can be used for the determination of the inductive influence of HV or EHV cable lines during normal operating conditions. In that case, it is only necessary to know the maximal load currents and the spatial position of each phase conductor. Also, in that cases where the values of the equivalent screening factors are noticeably different for different phase conductors, this factor should be adopted in such a way that corresponds to the phase conductor with the lowest total screening effect. In this way the whole procedure is simplified while the obtained results are slightly on the side of larger safety. However, in the cases where the predicted inductive influence is estimated as higher than the prescribed one, the protective measure is mandatory. Then it is necessary to determine the relevant parameters of the equivalent screening conductor.

Finally, the flowchart describing the whole here-presented calculation procedure for the determination of the induced voltages and currents are shown in Fig. 7.10.

Difficulties in the application of this methodological version arise from the fact that it is impossible to accurately determine potential U_{im} as well as impedances $Z_{C'}$ and $Z_{D'}$, by measurements. The reason for that is the impossibility of finding the so-called remote ground in urban conditions and the inductive influence of the surrounding metal installations. Therefore this methodology version is more suitable and more accurate for application in suburban areas.

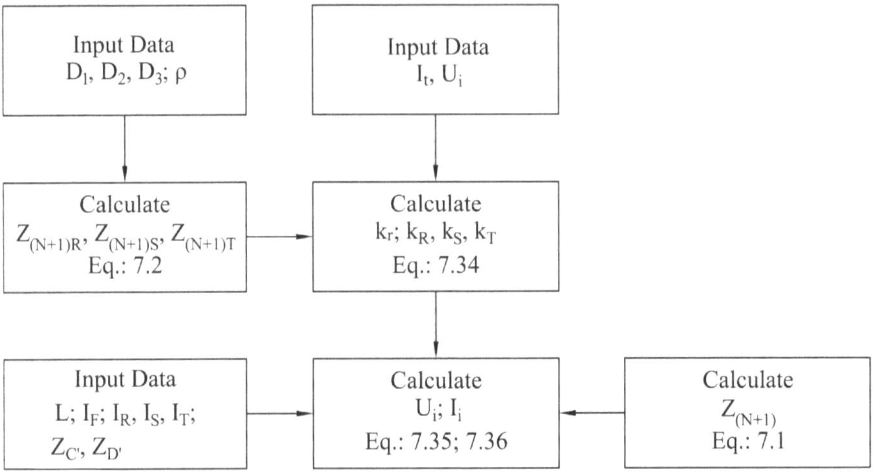

Figure 7.10 Flow diagram of necessary calculations.

7.3.3 Relevant parameters of equivalent screening conductor

When the estimated value of the induced voltage and/or current is larger than a tolerable one, the measure for their mitigation is indispensable. Such a measure usually implies laying the so-called compensating conductor [19]. For its practical realization, it is necessary to determine its constructive characteristics so that this conductor corresponds to its purpose and reduces the induced voltage and current to the allowed values. However, for this conductor not to be oversized, it is necessary before that to determine relevant parameters of the introduced equivalent shielding conductor.

With the aim of simplifying the problem, let us imagine an equivalent screening conductor substituting, from the standpoint of the screening effect, all surrounding metal installations, including all neutral (screening) conductors of the inducing line. The problem of determining such a conductor in the case considered in Section 7.2 has become solvable when it is assumed that this equivalent conductor has a cylindrical form and is placed around the considered phase conductor along the entire inducing line length [6]. However, in this case, it is necessary to assume that such a conductor is placed around the induced (designed) metal installation only along the section of its exposure. Also, it is assumed that this conductor is perfectly grounded at both ends. Under these assumptions the induced metal installation and its equivalent screening conductor can be presented as shown in Fig. 7.11.

On the basis of the equivalent conductor imagined in the described way, the corresponding equivalent circuit of one inducing conductor (e.g., phase conductor T with the current I_t) and of the previously mentioned provisory cable line becomes simple as shown in Fig. 7.12.

The used notation has the following meaning: $U_{(N+1)T}$ is the voltage induced by current I_t in the provisory cable line; $U_{(N+1)eq}$ is the voltage induced by current I_i in

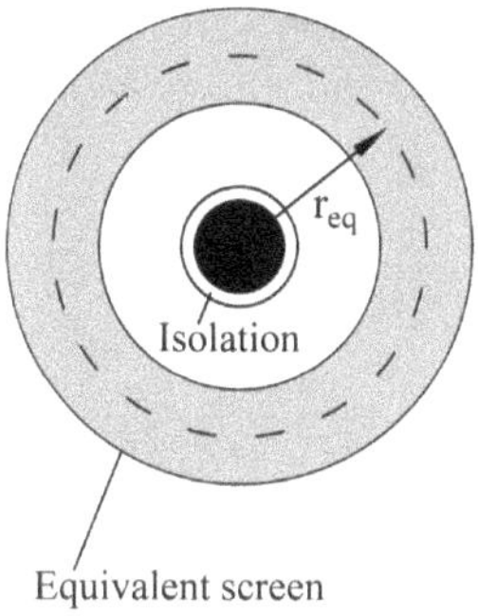

Figure 7.11 Cross section of the induced installation and the equivalent screening conductor.

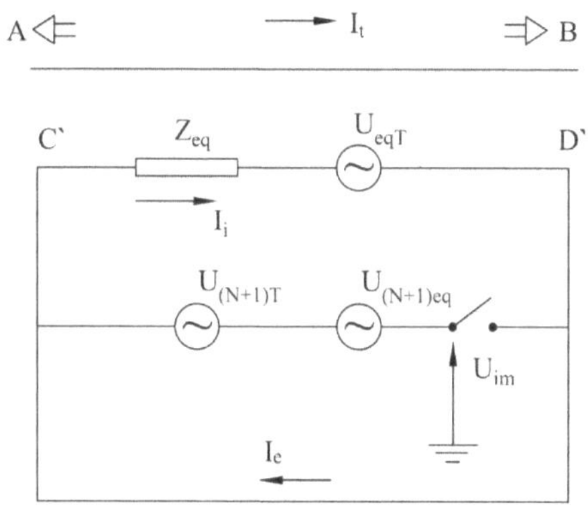

Figure 7.12 The simplified equivalent circuit of the provisory cable line.

the provisory cable line; U_{eqT} is the voltage induced by current I_t in the equivalent screening conductor; U_{im} is the measured value of the voltage induced along the provisory cable line (V); I_t is the simulated ground-fault current; I_i is the current induced in the equivalent screening conductor; Z_{eq} is the self-impedance of the equivalent screening conductor; and Z_{eqT} is the mutual impedance between the equivalent screening conductor and the phase conductor T (Ω).

The relevant parameters of the introduced equivalent screening conductor can be determined under the assumption that the voltage induced in the provisory cable line remains unchanged by introducing this conductor. Based on the equivalent circuit in Fig. 7.12, this condition can be expressed by the following system of equations:

$$\begin{aligned} &1.\ Z_{(N+1)T}I_t + Z_{eq}I_i = 0 \\ &2.\ Z_{(N+1)T}I_t + Z_{(N+1)eq}I_i = U_{im} \end{aligned} \tag{7.37}$$

where Z_{eq} is the self-impedance of the equivalent screening conductor (Ω).

For the imagined physical appearance of the equivalent screening conductor, impedances Z_{eq}, $Z_{(N+1)T}$, and Z_{eqT} are in advance known and, according to Ref. [20], per unit length determined by

$$Z'_{eq} = R'_{eq} + \omega_t \frac{\mu_0}{8} + j\omega_t \frac{\mu_0}{2\pi} \ln\frac{\delta}{r_{eq}}, \quad \Omega/\text{km}, \tag{7.38}$$

$$Z'_{(N+1)eq} = Z'_{eqT} = \omega_t \frac{\mu_0}{8} + j\omega_t \frac{\mu_0}{2\pi} \ln\frac{\delta}{r_{eq}}, \quad \Omega/\text{km}, \tag{7.39}$$

where R'_{eq} is the longitudinal resistance of the equivalent screening conductor (Ω/km), and r_{eq} is the mean radius of the cylinder representing the equivalent screening conductor (m).

Since voltage U_{im} can be determined by measurement, and since the impedances $Z_{(N+1)T}$ and Z_{eqT} can be determined by calculations, these quantities can be considered known and the previously defined condition, Eq. (7.37), can be expressed in the following, simpler form:

$$\frac{Z_{eq}}{Z_{(N+1)eq}} = \frac{U_i}{U_i - U_{im}}, \tag{7.40}$$

where U_i represents the voltage that would be induced under the assumption that the screening conductors do not exist (V).

Since expression (7.40) gives a relationship between complex quantities, it can be represented by the following system of two equations:

$$\begin{aligned} \text{Re}\left\{(U_i - U_{im})Z'_{eq}\right\} &= \text{Re}\left(U_i Z'_{(N+1)eq}\right) \\ \text{Im}\left\{(U_i - U_{im})Z'_{eq}\right\} &= \text{Im}\left(U_i Z'_{(N+1)eq}\right). \end{aligned} \tag{7.41}$$

where Re and Im are the real and imaginary parts of Eq. (7.40).

Then, the relevant parameters of the introduced equivalent screening conductor (R'_{eq} and r_{eq}), based on Eqs. (7.33), (7.34), and Eq. (7.36), can be determined by

$$R'_{\text{eq}} = \text{Re}\left\{\frac{U_i}{U_i - U_{im}} Z'_{(N+1)eq}\right\} - \omega_t \frac{\mu_0}{8}, \tag{7.42}$$

$$X'_{\text{eq}} = \text{Im}\left\{\frac{U_i}{U_i - U_{im}} Z'_{(N+1)eq}\right\}, \tag{7.43}$$

$$r_{eq} = \frac{\delta_t}{e^{2\pi X'_{eq}/\omega_t \mu_0}}, \tag{7.44}$$

where R'_{eq} (X'_{eq}) is the real (imaginary) part of the self-impedance of the equivalent screening conductor (Ω/km).

After determining the relevant parameters of the introduced equivalent screening conductor (R'_{eq} and r_{eq}), Eqs. (7.38) and (7.39) can be used to obtain Z_{eq}, Z_{eqT}, and $Z_{(N+1)eq}$, as well as I_i and I_e (Fig. 7.12). Since voltages U_i and U_{im} are the phasor quantities, to obtain more accurate results, it is also necessary to determine the phase angle between them. This can be done by measuring the phase angle between U_{im} and I_t and by the corresponding calculation based on the equivalent circuit of Fig. 7.12.

7.3.4 Determination of compensating conductor

When the relevant parameters of the equivalent screening conductor are determined, it becomes possible to determine the necessary constructive parameters of the compensating conductor, as well. This conductor (usually bared copper wire) is laid right next to the induced installation along the whole length of this installation and is grounded at both ends, usually via the buildings' foundation armatures. Thus its spatial position is a priori known, whereas the grounding impedances at its ends can be determined by measurements or by calculations that take into account the grounding effects of the compensating conductor itself.

In order for this conductor to meet its purpose, it is necessary that the current circulating through it be sufficiently large. In other words, the total impedance of the electrical circuit that forms this conductor and the ground as a return path needs to be small enough. Thus its constructive characteristics should be determined in such a way that the voltage induced along the exposed section does not exceed a certain limit value. The constructive characteristics of this conductor that meet the set criterion can be determined by using the equivalent circuit in Fig. 7.13.

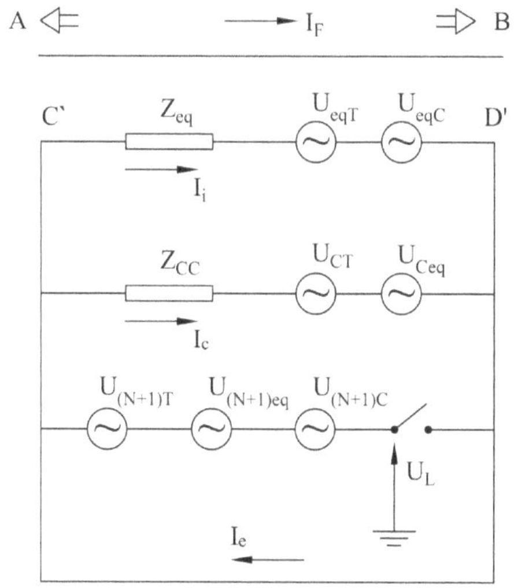

Figure 7.13 Equivalent circuit of the compensating conductor.

The used notation has the following meaning: $U_{(N+1)T}$ is the voltage induced by current I_F in the designed installation (V); $U_{(N+1)eq}$ is the voltage induced by current I_i in the designed installation (V); $U_{(N+1)C}$ is the voltage induced by current I_c in the designed installation (V); U_{eqT} is the voltage induced by current I_F in the equivalent screening conductor (V); U_{CT} is the voltage induced by current I_F in the compensating conductor (V); U_{eqC} is the voltage induced by current I_c in the equivalent screening conductor (V); U_{Ceq} is the voltage induced by current I_i in the compensating conductor (V); U_L is the limit allowable value of the induced voltage in the designed installation (V); I_F is the anticipated maximal ground-fault current (A); I_i is the current induced in the equivalent screening conductor (A); I_c is the current induced in the compensating conductor (A); Z_{eq} is the self-impedance of the equivalent screening conductor of the exposure section (Ω); and Z_{CC} is the total impedance of the compensating circuit (Ω).

The total impedance of the compensating circuit is determined by

$$Z_{CC} = Z_C + Z_{C'} + Z_{D'}, \tag{7.45}$$

where Z_C is the self-impedance of the compensating circuit (Ω); and $Z_{C'}$ ($Z_{D'}$) are the grounding impedances at the ends of the exposure section.

Based on the equivalent circuit in Fig. 7.13 one can write the following system of equations:

$$\begin{aligned}
&1.\ Z_{eq}I_i + Z_{eqT}I_F + Z_{eqC}I_c = 0\\
&2.\ Z_{CC}I_c + Z_{CT}I_F + Z_{Ceq}I_i = 0\\
&3.\ Z_{(N+1)T}I_F + Z_{(n+1)eq}I_i + Z_{(N+1)C}I_c = U_L
\end{aligned} \tag{7.46}$$

Then based on the known relevant parameters of the equivalent screening conductor and spatial position of the compensating conductor, we can determine the impedance Z_{eq} and all mutual impedances in the system of Eq. (7.46). Since we can also determine by calculations or by measurements the grounding impedances $Z_{C'}$ and $Z_{D'}$, as well as since the anticipated ground-fault current is known based on separate studies, the only unknown quantities in a given system of equations are currents I_i and I_c, as well as impedance Z_C. Thus by solving this system of equations, one obtains impedance Z_C and based on it and corresponding analytical expression for the self-impedance for such a conductor, the necessary constructive characteristics of the compensating conductor, R'_C, and r_c. Based on determined R'_C, and r_c, one can adopt the corresponding compensating conductor that should be installed together with the designed installation to reduce the induced voltage to the allowed values and ensure the necessary safety conditions and/or equipment integrity. By taking into account the screening effects of all known and unknown surrounding metal installations, this conductor can be selected in accordance with the actual needs.

Certainly, an analogous procedure can be used for the determination of the compensating conductor capable of protecting the designed metal installation from the too high inductive influence of HV or EHV cable lines during normal operating conditions. It is only necessary to have data concerning the maximal load currents and the spatial position of each phase conductor. In the cases when the relevant parameters of the equivalent screening conductor are noticeably different for different phase conductors, those parameters should be adopted in such a way that corresponds to the phase conductor of the lowest screening effect of surrounding metal installations. In this way the whole procedure is simplified and the obtained results are slightly on the side of the enlarged safety.

If the designed metal installation has its screen, the presented procedure can be used to determine its relevant parameters so that it ensures the desired mitigation level of induced voltages and currents without using a specially and separately constructed compensation conductor. Finally, the screening effect achieved by the compensating conductor can be checked by applying the test circuit shown in Fig. 7.9 to the induced installation when constructed.

7.4 Quantitative analysis

7.4.1 Methodology verification

For this purpose, let us assume one 110-kV distribution line consisting of three single-core cables laid in a trefoil formation along the whole line. The phase conductors of these cables are made of aluminum having a cross section of 1000 mm^2, while their metal sheaths are made of copper strings having a total cross section of 95 mm^2, and the mean diameter is equal to 91 mm. Also, let us assume that the equivalent soil resistivity of the area through which this line is laid is 30 Ω m.

Then, to verify the presented methodology, let us assume that a ground-fault current of 20 kA flows through one of the line cables, while the metal sheath of one of the other two cables is not grounded at one of the line ends. Also, we will assume that there are no other metal installations in the surrounding area. According to these assumptions, this cable line can be presented by the equivalent circuit shown in Fig. 7.14.

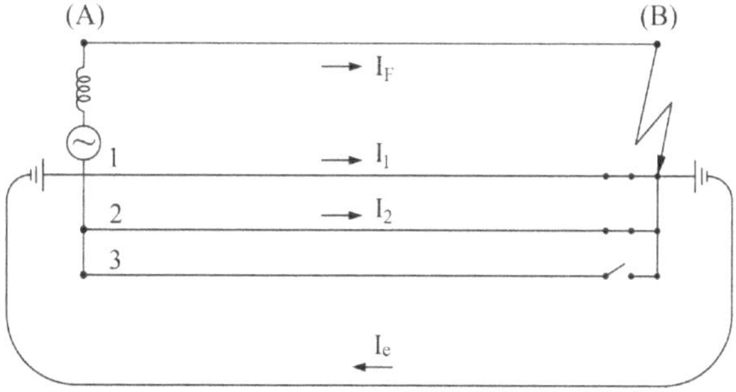

Figure 7.14 The equivalent circuit of the assumed cable line.

By using the existing method based on the known all necessary data, the given equivalent circuit, and the corresponding closed system of equations (e.g., Ref. [23]), it can be shown that the voltages induced in the metal sheath having one end not grounded take the following values: $U_i = (986 + j10846)$ and $U_{im} = (1735 + j276)$ V/km. Then, by using these results as input data for the calculation procedure presented here, it can be shown that the relevant parameters of the equivalent screening conductor substituting two cable line screens grounded at both ends are $R'_{eq} = 0.0936\ \Omega$/km, $r_{eq} = 66.1$ mm. Finally, by using these values, it is not difficult to show that the resulting screening factor of the two cable screens in the considered case is

$$k_r = (0.0438 + j0.1678).$$

Finally, by using this value, the induced voltage has been obtained the value that is the same as the one obtained by using the existing method and the known data about the considered line. Thus it can be said that the accuracy of the here-presented method has been confirmed.

On the basis of the presented analytical expressions, one can see that for performing corresponding calculations, it is necessary to know the value of the equivalent resistivity of the surrounding soil. However, although there are several methods of measurement of soil resistivity, none of them is applicable in urban conditions. This is because the surfaces of urban areas are covered by buildings, streets, pavements, and many other permanently constructed objects, while belowground many known and unknown metal installations are already present and unavoidably influence the results of measurements. Thus we are forced to adopt an approximate value of the equivalent soil resistivity, based on the main geological characteristics of the surrounding area, and use it in the necessary calculations. However, here it is important to remind that the calculated quantities necessary for the application of here-presented methodology are only slightly dependent on the equivalent soil resistivity so that certain arbitrariness in determining the equivalent soil resistivity does not have any practical significance. It is sufficient to know that, within the whole range of actually possible values, the lowest one should be preferred because it gives the final result which is slightly on the side of the enlarged safety. Thus it can be said that the accuracy of the developed methodology depends mainly on the accuracy of the measurement of the voltage U_{im} and phase angle between this voltage and test current in the inducing line, I_t.

The methodological version presented in Sections 7.2 and 7.3 takes into account almost all relevant factors and parameters, including those that are unknown and those, influence of which is negligibly small. Thus the accuracy of the presented methodology depends mainly on the accuracy of the necessary test measurements. Some inaccuracies in measurements can appear only as a consequence of the inductive influence of the nearby power distribution lines. This influence can be efficiently avoided by using the test current of somewhat higher frequency which can easily be discriminated from the omnipresent power frequency. The error introduced in such a way is small and its influence on the final calculation results is such that they are slightly on the side of the enlarged safety.

7.4.2 *Screening effects of the surrounding metal installations*

With the aim of obtaining an insight into the magnitude of the considered screening effect of the surrounding metal installations in the case of the previously defined cable line, let us assume that this line is laid in a typical urban environment. The desired insight into the magnitude of the considered effect we will obtain on the basis of the results of the experimental investigations performed in the 110-kV distribution network of Beograd [1,2] and presented in Section 3.6. These investigations performed on two cable lines with the identical technical characteristics given in the previous section show that the relevant parameters of the equivalent screen substituting all surrounding metal installations are the following:

- in the case of the line passing through the lower degree of urbanization: $R'_{eq} = 0.08\ \Omega/\text{km}$, $r_{eq} = 12.3$ m and
- in the case of the line passing through the higher degree of urbanization: $R'_{eq} = 0.053\ \Omega/\text{km}$, $r_{eq} = 8.1$ m.

At this, it was adopted that the equivalent soil resistivity of the area through which these lines are laid is 30 Ω m. As it is known, this is a relatively low value of the equivalent soil resistivity, so that the considered screening effect is less pronounced here than it would be in the conditions of higher specific soil resistivity.

Based on these data and here-presented analytical expressions, it is not difficult to show that in the case of the line with cross-bonding, the induced voltage in any of the surrounding metal installations during normal operating conditions is smaller by 70% in the case of the lower and by 81% in the case of the higher degree of urbanization. Underground-fault conditions the induced voltages in any of the surrounding metal installations are smaller by 54% in the case of the area of the lower and by 67% in the case of the area of the higher degree of urbanization.

Also, let us assume that the maximal anticipated ground-fault current in the supplied substation is 20 kA and that the feeding line and the induced pipeline are laid under the same pavement at a separation of 1 m from the nearest phase conductor. Then, if the surrounding metal installations are disregarded, one obtains that the voltage induced in the pipeline is 942 V/km. However, if one takes into account the screening effects of the surrounding metal installations, this value is significantly smaller, that is, 434 V/km for the lower, and 311 V/km for the higher degree of urbanization. Finally, if one assumes that the pipeline is at a distance of 20 m from the nearest phase conductor, the corresponding induced voltages, under identical all other conditions, would be 500, 230, and 165 V/km, respectively.

Then, let us assume that the considered line is without cross-bonding and that its length is equal to 4 km (an approximately average length in the 110-kV distribution network of Beograd). On the basis of these assumptions, the maximum stress voltages to which the cable screen coating will be exposed at the critical place, in the middle of the line since the screens are earthed at both line ends is

- without taking into account the surrounding metal installations: $(3470 + j552)$ V,
- in the case of the lower degree of urbanization: $(799 + j1413)$ V, and
- in the case of a higher degree of urbanization: $(578 + j1088)$ V.

The presented results show that the favorable screening effect of the surrounding metal installations is very pronounced. Because of that it is very useful to take it into consideration when one evaluates the inductive influence of HV and EHV lines passing through urban and/or suburban areas. Also, the presented results show that the considered effect significantly depends on the degree of urbanization and it is higher where the urbanization degree is higher, that is, where this favorable effect is more needed. Finally, these results significantly change our earlier perception concerning the magnitude of this problem. By taking into account the screening effect of the surrounding metal installations, we put the whole problem in the realistic frameworks and avoid in many critical cases the unnecessary expenditures for the protection measures against too high induced voltages and currents.

7.5 General observations

The inductive influence of HV and EHV lines on different nearby metal installations had been the subject of voluminous research for at least six last decades. Hundreds of research papers, including national and international directives (e.g.: CCITT, International Telegraph and Telethone Consultative Committee), were published to enable determining this influence in all practically possible conditions and circumstances and undertaking corresponding protective measures against too high induced voltages and currents. However, the formerly known methodologies are applicable only in a limited number of cases, where a scenario of the right-of-way is not too complicated and where all relevant factors and parameters are known. Even in such conditions and circumstances, because of the complexity of the practical problems, a necessary solution frequently was not possible without using modern computers. Also, none of the formerly developed methodologies enables solving the problem in the cases when it appears in urban or suburban areas where many known and unknown surrounding metal installations act as screening conductors and as such reduce the induced voltages and currents. Thus because of many relevant, but uncertain and completely unknown data about the surrounding metal installations in these conditions and circumstances, the practical problems were usually solved by using too rough neglect that the surrounding metal installations do not exist. As a final consequence, expensive protective measures against too high induced voltages and currents can be also applied in cases where they are not indispensable.

The presented methodology enables the correct prediction of the voltages and currents that can appear for any metal installation that will be in an urban or suburban area under the inductive influence of an HV or EHV line. The main advantage of the methodology in comparison to the earlier known ones is that it is such that compensates the deficiency of an arbitrarily large number of relevant, but unknown data concerning the surrounding metal installations present in urban and suburban areas (their number, spatial disposition, and technical characteristic). As such it does not need the application of modern computers and is very easy and simple for practical application in all practically possible cases. Besides, this methodology is more suitable for application in cases that were so far solved by some earlier known and accepted methodologies, as well.

The presented methodology enables the determination of the screening effects of many different, known and unknown, surrounding metal installations, reducing the magnitude of the inductive influence of HV and EHV lines passing through urban and/or suburban areas. In each concrete case, these installations can be completely represented by only one equivalent screening conductor belonging to the inducing line or induced installation. This is enabled by defining the necessary test measurements and by the corresponding calculation procedure based on the results of these measurements. By determining the relevant parameters of the introduced equivalent screening conductor, each practical situation, irrespective of its complexity and the amount of relevant but uncertain and unknown data, is replaced by the simplest case from the standpoint of the number of inductively coupled circuits. From the mathematical point of view, it is interesting to note that here-presented methodology enables a solution in the cases, where, because of the deficiency of a large number of relevant data, the necessary system of equations enabling the application of the previously known methodologies cannot be completed.

Since the ground-fault current in an HV distribution network can be more than 20 kA, induced dangerous and harmful voltages and currents can appear in different metal installations that pass along the same streets as an HV or EHV line. On the basis of the given quantitative analysis, it is shown that the potentially dangerous and harmful influence of HV or EHV lines upon any surrounding installation seen individually is smaller if the degree of urbanization is larger, that is, if the number of surrounding metal installations is larger.

In unsettled areas due to strict environmental regulations, HV or EHV lines and other types of metal installations are placed in such a way that they share common corridors and as a result, too high voltages and currents can be induced in nearby metal installations, particularly in cases with a long length of exposure and/or a high level of ground-fault currents. In such conditions and circumstances, the presented methodology enables solving the problem without the need for a collection of a large number of relevant data and application of the complex mathematical model based on the possibilities of modern computers. All necessary calculations can be performed by using handheld calculators only.

The presented methodology enables determining the voltages and currents induced in any of the metal installations surrounding an HV or EHV line in any (urban, suburban, or uninhabited) surroundings. The maximal induced voltages and currents can be determined during normal operation, as well as during fault conditions of these lines. In both cases, it is sufficient to know the relevant constructive characteristics of the considered nearby installation and its spatial position with respect to the inducing line.

With full knowledge of the electromagnetic induction, utilities can be allocated acceptable right-of-ways for their power lines, industrial and commercial buildings can be allocated on sufficient distances from the HV and EHV lines, and utilities can be given the good separation of their pipelines and railroads from these lines.

Bearing in mind all these facts, the presented methodology could be used as a foundation for a certain amendment of current versions of the corresponding technical standards and guides.

References

[1] Lj.M. Popović, Influence of metal installations surrounding the feeding cable line on the ground fault current distribution in supplied substations, IEEE Transactions on Power Delivery 23 (4) (2008) 2583–2590.
[2] Lj.M. Popović, Testing and evaluating grounding systems for substations located in urban areas, IET Generation Transmission and Distribution 5 (2) (2011) 231–238.
[3] Lj.M. Popović, Determination of the actual reduction factor of HV and MV cable line passing through urban and suburban area, Facta Universitatis-Series: Electronics and Energetics 27 (1) (2014) 25–39.
[4] Lj.M. Popović, Ground fault current distribution when a ground fault occurs in HV substations located in an urban areas, Progress In Electromagnetic Research B 59 (2014) 167–179.
[5] Lj.M. Popović, New method for determination of inductive influence of one or more power lines on any of exposed metal installations, International Journal of Engineering Research and Application 6 (2) (2016) 36–46.
[6] Lj.M. Popović, Inductive influence of HV cable lines in urban and suburban areas, Electric Power System Research 176 (2019) 105944.
[7] R.J. Carson, Wave propagation in overhead wires with ground return, Bell Systems Technical Journal 5 (4) (1926) 539–554.
[8] F. Polackzek, On the field produced by an infinitely long wire carring alternating current, Electrische Nachricten Technik III (9) (1926) 339–359.
[9] E. Sunde, Earth Conduction Effect in Transmission Systems, Van Nostrand Company, New York, 1949.
[10] J. Pohl, Influence of high-voltage over-head lines on covered pipelines, CIGRE, Paper No. 326, 1966.
[11] F. Dawalibi, A. Pinho, Computerized analysis of power systems and pipelines proximity effects, IEEE Transactions on Power Delivery 1 (2) (1986) 40–48.
[12] S.T. Sobral, W. Caustingheiras, V.S. Costa, D. Mukhedkar, Interference between faulted power circuits and communication circuits or pipelines—simplification using the decoupled method, IEEE Transactions on Power Delivery 6 (4) (1991) 1599–1606.
[13] K.J. Satsios, D.P. Labridis, P.S. Dokopoulos, The influence of multi-layer ground on electromagnetic field of an overhead power transmission line in the presence of buried conductors, Archiv fur Electrotechnik 80 (5) (1997) 343–348.
[14] G. Djogo, M.M.A. Salama, Calculation of inductive coupling from power lines to multiple pipelines and buried conductors, Electric Power System Research 41 (1) (1997) 75–84.
[15] G.C. Christoforidis, D.P. Labridis, P.S. Dokopoulos, Inductive influence on pipeline buried in multilayer soil due to magnetic fields from nearby faulted power lines, IEEE Transactions on Electromagnetic Compatibility 47 (2) (2005) 254–262.
[16] L. Bortels, J. Docaninck, J.I. Pineiro, C. Munteanu, V. Topa, A general applicable model for AC predictive and mitigation techniques for pipeline networks induced by HV power lines, IEEE Transactions on Power Delivery 21 (1) (2006) 210–217.

[17] D.A. Tsiamitros, G.K. Papagiannis, D.P. Labridis, P.S. Dokopoulos, Earth return path impedances of underground cables for the two-layers earth case, IEEE Transaction on Power Delivery 20 (3) (2005) 2174–2181.
[18] D.A. Tsiamitros, G.K. Papagiannis, P.S. Dokopoulos, Homogenous earth approximation of two-layer earth structures. A. Equivalent resistivity approach, IEEE Transactions on Power Delivery 22 (1) (2007) 658–666.
[19] CCITT directives concerning the protection of telecommunication lines again harmful effects from electric and electrified railway lines, ITU, Geneva, 1990 (project report).
[20] Int. Std., ref. CEI/IEC 60908-3, (Ed.) 0.3 2009-3, short-circuits currents in three-phase A.C. systems—Part 3: Currents during two separate simultaneous line-to-earth short circuits and partial short–circuits following through earth.
[21] IEEE guide for bonding shields and sheaths of single-conductor power cables rated 5 kV through 500 kV, IEEE Standard 575-2014, 2014.
[22] NACE Standard Recommended Practice SPO177-2007, Mitigation of alternating current and lightning effects on metallic structure and corrosion control systems, 2007.
[23] Lj.M. Popović, Practical Methods for Analysis and Design of HV Installation Grounding Systems, Academic Press, Elsevier, Oxford, 2018.
[24] Lj.M. Popović, Determination of the reduction factor for feeding cable lines consisting of three single-core cables, IEEE Transactions on Power Delivery 18 (3) (2003) 736–743.

Transfer characteristics of power lines passing through urban areas

8

8.1 Introduction

8.1.1 Historical development of power systems

Electric power is universally utilized and required in the production of commodities and services. It is provided almost at each place where people live at any time during the day, 7 days a week, year after year. Without it, human life in the nowadays world would be simply unimaginable. The influence of this permanent delivery on the achieved quality of modern human life is so high that this period of our civilization is frequently called the era of electrification. However, this welfare is mostly taken for granted, and no like something that is the result of the great devotion and continuous research work of many generations of engineers and scientists and above all extremely gifted individuals among them. Because of that let us consider the electric-power system development starting from the very beginning of its appearance to the present days.

Today's AC (alternating current) electric power system has been evolving since 1896 on the basis of Nikola Tesla's patents comprising a comprehensive system of generation, transmission, transformation, poly-phase motors, and light [1]. Generally speaking, Tesla's innovations enabled that electrical energy is obtained at places where energy is abundant and used at places where there are needs for different kinds of useful energy. Since this concept is such that it is not possible to add or subtract something to it, it remained essentially unchanged until the present days. As such this concept has had a decisive influence on the goals and main directions of electric-power systems development, advancement, and magnification throughout the period longer than a hundred years. During this whole period, Tesla's ACs were used as a well-rounded concept of production, transmission, and consumption of electricity, which is dominant away even today all over the world.

The constant increase in electricity consumption and the increase in the power of individual consumers have led to a constant increase in the production, transmission, and distribution capacities of electric power systems. Besides, over time, there have been appeared demands for greater reliability in delivery and higher quality of delivered electricity.

Certainly, at the very beginning, electric-power systems have been constructed only as small local networks. For instance, the first one built in 1895 served for covering the needs of only one city, Buffalo (the United States). It was supplied by the energy produced by hydro turbines installed at 22 miles distant Niagara Falls power station [1]. Somewhat later, during the 20th century, they became larger and

Relevant Characteristics of Power Lines Passing through Urban Areas. DOI: https://doi.org/10.1016/B978-0-323-91136-8.00005-0

larger and more and more mutually interconnected for economic and reliability reasons. Cheap electrical energy obtained from power networks initiated and enabled a rapid and massive industrial development over the entire world. All of this had the feedback effects that resulted the further growth of these networks to serve the industry but also for commercial and residential purposes.

By the 1960s, as a consequence of dynamic development, the electric-power networks of developed countries have become very large, highly interconnected, with thousands of generation power stations delivering power to major load centers via high-capacity transmission lines. On the other hand, these networks were characterized by the lack of flexibility and built-in intelligence, aging infrastructure and controls, relatively low reliability and efficiency, increased consumer cost of service issues, and environmental issues. They generally implement specific functions as individual applications. Supervisory control and data acquisition applications provided monitoring and control down to distribution feeder breakers and, in a few cases, beyond the feeder circuit breaker on the feeders themselves (outside the substation fence). Voltage and var controls are implemented at the individual (standalone) capacitor banks and regulators. Since the electric-power network at that time lacked a communication network, coordination between devices was handled via time-coordinated controller settings based on planning studies. Demand response systems were typically implemented on the basis of central controls with little coordination at the distribution level. The metering capabilities of these networks meant technical limitations on the degree to which price signals could be propagated through the system.

Through the last decades of the 20th century, growing demand led to increasing the number of power stations. In some areas the supply of electricity, especially at times of peak demands, could not keep up with these demands, resulting in poor power quality, including blackouts, power cuts, and brownouts. Also, electricity demand patterns were established: heating and air-conditioning led to daily peaks in demand that were met by an array of peaking power generators only be turned in for short periods each day. This situation, together with increased losses in the whole network, resulted in high costs to electricity companies, which were passed on in the form of increased tariffs.

8.1.2 Contemporary tendencies in utilization of high-voltage and extra high-voltage lines

At the beginning of the 21st century, opportunities to overcome the problems of limitations and costs of power networks were enabled by improvements in electronic communications and modern computer technology. New technologies have emerged in power networks from earlier attempts at using electronic control, metering, and monitoring. Smart meters add continuous communications so that monitoring can be done in real time. The networks in which these new technologies would be implemented to a full extent, in professional literature, are known under the term "smart grids." Many definitions can be used to explain this contemporary term, some functional, some technological, and some benefits-oriented. Thus let me mention only the following one hoping it is the most comprehensive one. That is an electric-power network whereby instantaneous coordination of many different functions is achieved economical and rational utilization of all available generation and transfer resources at each moment of normal operation as well as contingency conditions.

When a smart grid would be completely achieved, from such an electric-power network, one can expect the following advantages and conveniences:

- High reliability, because new technologies will improve fault detection and allow self-healing of the network without any intervention of technicians.
- Flexibility in network topology, because next-generation transmission and distribution infrastructure will be able to handle possible bidirectional energy flows, allowing for distributed generation such as photovoltaic panels mounted on buildings roofs, fuel cells, charging to/from the batteries of electric cars, wind turbines, pumped hydroelectric power, and other sources.
- Higher efficiency of numerous contributors to an overall improvement of the efficiency of energy infrastructure, particularly, including the relation: demand-side management. For example, turning off air conditioners during short-time spikes in electricity price. The overall effect is less redundancy in transmission and distribution lines, and greater utilization, leading to lower power prices.

On the basis of this short survey of the directions and goals of the developments of electric-power networks, it can be noted that many technical solutions and new technologies have been introduced into these networks from their first appearance to the present days. A power network is now probably the largest and the most complicated ever constructed technical system. Their worth is enormous and, only as an example, the power network of North America was estimated in 2004 to be US$800 billion [2]. However, power transfer capacities of the transmission and distribution lines are determined by using the calculation procedure that is practically the same from the beginning of electric-power networks' exploitation. Thus transmission and distribution lines during the whole period of their exploitations have had continuous ratings equal to the maximum planned load. However, in practice, the load on transmission or distribution lines may vary with time owing to variations of the load pattern of the system, and maximum values only occur for limited periods. Also, buried cables and the surrounding soil have a large thermal capacity, and steady-state conditions following a load change may not be reached for hours or days. Consequently, many cables could have the potential for increasing ratings over limited periods without exceeding the maximum design operating temperatures.

During the 1980s a cable system model using a digital microprocessor has been developed to exploit the transient transfer capacity of cables. This model includes a detailed thermal simulation of only individual extra high-voltage (EHV) cable lines and their environment [3]. One function of this model is to calculate the phase conductor temperature in real time using the equivalent thermal circuit of the cable fed with metered data from the cable route. Also, the recent deregulation process has raised interest in the thermal capability of conductors of overhead lines for maximizing their power transfer capacity [4]. In accordance with this, engineers have recognized the limitations of the conservative fixed ratings and the potential benefits of using dynamic thermal ratings that can be defined as a process of adjusting the thermal rating of power equipment in accordance with ambient conditions.

Also, in 2014 the method taking into consideration the favorable influence of surrounding metal installations on the transfer characteristics of power lines passing through urban and suburban areas is developed [5,6]. Since the surrounding metal installations remarkably reduce the inductive component of series impedance of HV and EHV lines, they significantly improve the power transfer capacity of these lines. Their attendance has effects similar to those that can be achieved by reactive energy compensation, that is, these installations reduce transmission line losses and increase the power transfer ability of HV and EHV transmission and distribution lines. This fact enables through the power-flow study optimum use of existing HV and EHV distribution networks, or in the other words to postpone the very expensive investments into the new HV and EHV cable lines. The performed quantitative analysis shows that the useful effects of surrounding metal installations on the power transfer capacity of the HV and EHV cable lines are remarkable [7].

The possibility of obtaining accurate values of HV and EHV power-distribution lines series impedance parameters is of prime importance to both the planning of the future development of HV and EHV distribution networks and the operation and protection of existing lines. Only data about series impedances of HV and EHV lines that correspond to the ones existing in practice enable correct both power-flow and fault studies in the networks to which these lines belong. In particular, without the actual values of these impedances, it is not possible either to determine the actual power transfer capacity of one HV or EHV distribution network. Also, errors in distribution-circuit impedance calculation can result in significant deviation in predicting fault current magnitudes, and thus in lack of proper protective devices' coordination.

The problem of accurate determination of the mentioned impedances appears because of the fact that HV and EHV cable lines are, almost without an exception, laid in urban or suburban areas, where there are many other, known and unknown, metal installations. The usage of common routes (mainly street sidewalks) for various supply networks (electricity, water, gas, oil, telecommunications, etc.) inevitably leads to their inductive coupling that affects the series impedance of these lines. Since these installations are mainly laid underground, many relevant data about them (constructive characteristic, mutual spatial position, including in some cases even and their total number) are uncertain or completely unknown. Also, the structure and spatial positions of these installations depend on many local factors and circumstances and, because of that, are different from case to case. Thus because of the many objective limitations and impossibilities concerning collecting necessary input data, the problem of determining the actual series impedance of distribution lines cannot be solved only by calculations. Also, the same or similar objective reasons prevent us to measure currents induced in the surrounding metal installations and an unbalanced current returning in the power system through the earth. Thus although the problem was well known in corresponding professional and scientific circles, it has been considered until recently as practically unsolvable. Finally, because of that, until recently designers and holders of HV and EHV distribution lines were forced, by determining their series impedances, to use the analytical expressions that completely disregard the existence of surrounding metal installations, for example [8,9]. The methodology that enables for the first time solving this problem was developed in Ref. [5], but before that, one more, also an important problem, had to be solved. That is the problem of actual ground-fault current distribution when a fault occurs in an HV substation located in an urban or suburban area.

The methodology that enables the compensation of deficiency of an arbitrary number of relevant but unknown data and solving the mentioned problem of ground-fault current distribution is developed in Refs. [10–12]. The solution was achieved by using the fact that a simulated ground-fault current in any of the phase conductors and its fraction in any of the line neutral conductors contain the cumulative effects of the inductive influence of all surrounding metal installations. The application of this methodology is based on a ground-fault simulation and test measurements of these currents in each concrete case of HV or EHV distribution line. The analytical part of the methodology uses these currents as known quantities and introduces in the development procedure one fictitious conductor that substitutes all surrounding metal installations, from the standpoint of their inductive influence. The final solution was achieved under the assumption that the imagined physical appearance and spatial position of the introduced equivalent conductor are such that it represents a cylinder that surrounds all conductors of the considered HV or EHV cable line along their entire length. The relevant parameters of this conductor are determined under the condition that its inductive influence on ground-fault current distribution is equal to the inductive influence of all surrounding metal installations. Based on the determined parameters of this equivalent conductor, it is not difficult by using the corresponding analytical expressions for the calculated reduction factor of a certain line to determine its actual reduction factor in each particular case, as is shown in Chapter 3, Actual reduction factor of power distribution lines.

Somewhat later the method was modified in such a manner that the introduced equivalent conductor substitutes, from standpoint of its inductive influence, all surrounding metal installations, including all neutral conductors of the considered line [13]. It is imagined as a cylinder around and along the phase conductor used for obtaining a simulated ground-fault current. As well as in the previous version, this conductor is completely determined by only two parameters. One of them is the mean radius of the cylinder representing this conductor, whereas the other one is the longitudinal resistivity of this conductor. After that, the developed methodology is adapted for solving an also actual problem in contemporary electrical engineering practice. That is the problem of the inductive influence of HV and EHV lines in urban and suburban areas [7].

The imagined physical appearance and spatial position of the equivalent conductor introduced in Refs. [11,12] are such that this conductor acts in fact as one additional neutral conductor for HV and EHV lines passing through urban and/or suburban areas. This fact has been utilized for developing the only until then known methodology that enables taking into account the favorable influences of surrounding metal installations on transfer characteristics of these lines [5,6]. However, the developed methodology, in that version, was not applicable in the cases of HV or EHV cable lines with cross-bonding, the most common case at all contemporary distribution HV and EHV cable lines [14].

The problem has been overcome by developing the corresponding version of the methodology of compensation of deficiency of relative but unknown data so that it enables solving the problem in the cases when HV and EHV cable lines are with applied cross-bonding [15]. The performed quantitative analysis shows that in the case of these cable lines, the favorable inductive influence of surrounding metal installations is significantly more pronounced than in the cases of cable lines with no applied cross-bonding. Such results have been, certainly, expected because it is obvious the application of cross-bonding intensifies the inductive influence of these power lines during normal operation on the surrounding metal installations. As a consequence of that endorsement, the feedback inductive influence of the surrounding metal installations on the inducing power line itself is enforced, as well.

8.2 Problem description

Bearing in mind all mentioned favorable effects of the inductive influence of surrounding metal installations, as well as the fact that laying a new HV or EHV cable line in the central, densely populated, areas of a modern town is a very expensive investment, one logical question arises by itself. Why the problem of determination of actual series impedances of the HV and EHV cable lines passing through urban and/or suburban areas was until recently without any solution?

Power-distribution networks in large urban agglomerations deliver electrical energy via many medium-voltage (MV), HV and, EHV lines passing through urban and/or suburban areas. For these power lines, such an environment means the immediate vicinity of many different metal installations typical for these areas. Because of their vicinity and mutual spatial disposition, determined mainly in accordance with the spatial disposition of streets, a fluctuating magnetic field existing around and along each power line induces voltages and currents in each of the surrounding metal installations. In this way, these installations spontaneously all together form one very complex electrical circuit around and along each HV or EHV line. Through the external electrodes of the grounding system of HV substations at the ends of these lines, all these metal installations, including line neutral conductors, are conductively firmly coupled. Because of all this, metal installations surrounding distribution lines spontaneously take over the function of additional neutral conductors of these lines. As such they have a certain returnable inductive influence on these lines by changing their transfer characteristics in normal operating conditions as well as during a ground fault.

The unbalanced current during normal operating conditions appears as a consequence of the fact that power-distribution systems consist of single-phase, two-phase, and/or untransposed three-phase lines serving unbalanced loads. Thus under normal operating conditions in a distribution network, an unbalanced current in the phase conductors leaves and returns into the power system through the earth and all neutral conductor(s), including the surrounding metal installations grounded at both ends. Its value is certainly significantly smaller compared to that existing during a ground fault, but its relative distribution between the earth and all available metal paths remains the same as in the case of a ground fault current, the case considered in Chapter 3, Actual reduction factor of power distribution lines. This is, certainly, because of the fact that in both cases these currents spontaneously form underground the same electrical circuit that is consisted mainly of the elements with constant parameters, that is, with parameters that are not dependent on the current passing through these elements.

Analytical expressions for the self and mutual impedances of the conductors having a common return path through the earth have been derived under the assumption that the power line is laid in a homogeneous soil of a resistivity equal to the equivalent resistivity of the normally heterogeneous (multilayer with each layer having a different resistivity) soil. Thus it can be said that the problem of determination of series impedance HV and EHV lines passing through urban areas has been, looked in principle, solvable for a long time (Chapter 1, Introductory considerations). However, it was a long time considered one unsolvable because collecting many relevant data concerning the metal installations in the vicinity of an HV or EHV in urban conditions is practically impossible.

Many metallic installations of different basic functions, like sheaths of different cable lines, neutral conductors of the low-voltage (LV) network, steel water pipes, reinforced concrete of building foundations, are situated nearby each HV or EHV cable line. Some of them are not in direct contact with the earth, while some are in direct and continuous contact with the earth on the whole their length. Also, they are interconnected in many places (mainly in each building of a contemporary town), and their spatial disposition and topology in each concrete case depend on many local conditions and circumstances. Besides, most of them are laid under the sidewalk along the streets so that much relevant data about them cannot be visually determined or verified.

The usage of common routes (mainly street sidewalks) for various supply networks (electricity, water, gas, oil, telecommunications, etc.) unavoidably leads to their mutual inductive coupling. Because of that the structure and topology of a network formed by numerous underground metal installations typical for urban areas are almost always very complex and many relevant data about them are uncertain and in many bases completely unknown. Thus because of these objective limitations and impossibilities of direct insight into these installations, input data necessary for calculations of series impedances of HV or EHV lines are mainly unreachable and sheltered. As a consequence of that the existence of surrounding metal installations is completely ignored in the calculation of the series impedance of power lines passing through urban and/or suburban areas, for example [8]. Thus the considered problem can be briefly defined in the following manner: how one can take into account the influence of surrounding metal installations at determining the series impedance of HV and EHV cable lines passing through urban and/or suburban areas?

Thus the problem appears because many of the relevant data concerning the surrounding metal installations are uncertain or completely unknown. The only that is surely known about them is most frequently the following:

- their structure and topology (spatial disposition) are different from case to case,
- some of them are in direct and continuous contact with the earth, whereas some others are grounded only at certain places, mainly in each building of the contemporary urban settlements,
- they are, through metal sheaths of the outgoing MV cable line, conductively connected to the substation grounding electrodes at the ends of each HV or EHV cable line, by the strict application terra-neutral (TN) earthing system in the LV network and consumer installations.

The fact that the structure and topology of the surrounding metal installations depend on many local conditions and circumstances also contributed to the impression that the problem is unsolvable. It practically means that for obtaining actual series impedance of power lines passing through urban and/or suburban areas, each of them should be considered individually, that is, as a case for itself. To this should certainly be added a great variety of degrees of urbanization and variety infrastructure objects in different parts of modern urban settlements. Thus for solving the considered problem was necessary the methodology that compensates for the deficiency of an arbitrarily large number of relevant but unknown data [5–7].

8.3 Complete equivalent circuit of an high-voltage or extra high-voltage cable line

The metal installations surrounding an HV or EHV line have different basic functions but since many of them are grounded at least at their ends, they also act as additional neutral conductors of these lines. According to this, each HV cable line passing through urban and/or suburban areas represents in reality a very complex electrical circuit with many conductively and inductively coupled elements. Assuming that the total number of surrounding metal installations is known such a cable line, in the general case, can be presented by the equivalent circuit shown in Fig. 8.1.

The used notation has the following meaning: A and B are the substations at the line ends; U_R, U_S, U_T, and I_R, I_S, I_T are the voltage sources and currents in the line phase conductors of the assumed cable line; U_{nR}, U_{nS}, U_{nT}, ..., U_{nN} are the voltages induced in an arbitrary, nth, circuit (metal installation) by the current in each of the surrounding circuit formed by line conductors or by surrounding metal installations; I_1, I_2, I_3 are the currents induced in the cable sheaths; I_e is the current passing solely through the earth; I_4, I_5, ..., I_n, ..., I_N are the currents induced in the individual surrounding metal installations; Z_{ph} is the self-impedances of the phase conductors; Z_1, Z_2, and Z_3 are the self-impedances of the cable sheaths ($Z_1 = Z_2 = Z_3$); Z_4, Z_5, Z_6, ..., Z_N are the self-impedances of the individual surrounding metal installations; Z_A (Z_B) is the grounding system impedance in substation A (B); N is the total number of

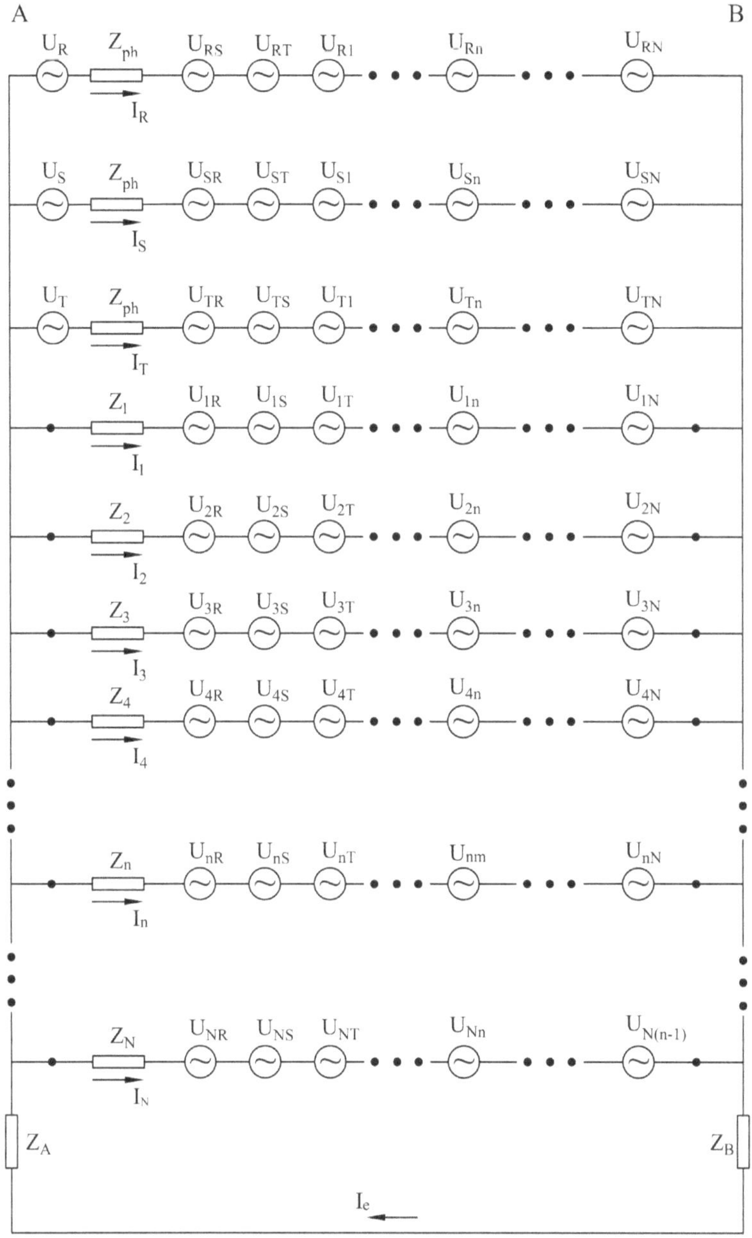

Figure 8.1 Complete equivalent circuit belonging to an HV cable line. *HV*, High voltage.

surrounding metal installations enlarged by the number of the HV cable line sheaths (This number, seen theoretically, can be arbitrarily large).

As can be seen, the presented equivalent circuit involves all self- and mutual impedances of all cable line conductors and all surrounding metal installations, as well as all induced voltages. The self-impedance of an arbitrary, nth, surrounding metal installation is, according to, for example [9], determined by

$$Z'_n = R'_n + \omega\frac{\mu_0}{8} + j\omega\frac{\mu_0}{2\pi}\left(\frac{\mu_r}{4} + \ln\frac{\delta}{r_n}\right), \quad \Omega/\text{km}, \tag{8.1}$$

whereas mutual impedance between two arbitrary, nth and mth, surrounding metal installations is determined by

$$Z'_{nm} = \omega\frac{\mu_0}{8} + j\omega\frac{\mu_0}{2\pi}\ln\frac{\delta}{d_{nm}}, \quad \Omega/\text{km}; \quad m \neq n \tag{8.2}$$

where R'_n is the longitudinal resistance of an arbitrary, nth, surrounding metal installation, Ω/km; r_n is the mean radius of an arbitrary, nth, surrounding metal installation (m); d_{nm} is the distance between two arbitrary, nth and mth, surrounding metal installations (m); ω is the angular power frequency:$2\pi f$; μ_0 is the magnetic permeability of vacuum: $4\pi \times 10^{-7}$ Vs/Am; μ_r is the relative magnetic permeability of the installation material; δ is the equivalent earth penetration depth.

The equivalent earth penetration depth, δ, is determined by

$$\delta = 658\sqrt{\frac{\rho}{f}}\ (\text{m}), \tag{8.3}$$

where ρ is the equivalent soil resistivity along and around the cable line (Ωm); and f is the power frequency.

Current in an arbitrary surrounding installation I_n (Fig. 8.1) induces in an also arbitrary (mth) surrounding installation a voltage, U_{mn}, which is determined by

$$U_{mn} = -Z'_{nm}I_n, \quad \text{V/km}, \quad m \neq n \tag{8.4}$$

where Z_{nm} is the mutual impedance between the mth and nth surrounding metal installation.

Also, for further considerations, it is necessary to mention that the current directions shown in Fig. 8.1 are taken arbitrarily.

By using the equivalent circuit shown in Fig. 8.1, one can write for each induced current from I_1 to I_N by the one equation and obtain a system of N independent equations. Then, based on the known values of the phase voltages (U_R, U_S, and U_T) and the known values of all self- and mutual impedances, it is possible to determine each of the induced currents by solving this system of equations. However, it is sufficient that only one of the self- or mutual impedances in the presented complete equivalent circuit is unknown and none of the induced currents can be determined and the whole problem seems unsolvable. Thus because of the numerous relevant but unknown data concerning the surrounding metal installations, even their total number, in many cases, such an approach is not practically possible.

Also, based on the equivalent circuit in Fig. 8.1, it is not difficult to see that each induced current in the surrounding conductors (cable line screens and surrounding metal installations) appears as a consequence of the inductive influence of the load currents through the line phase conductors and the currents induced in all other surrounding conductors. Thus each of the presented currents cumulatively involves the inductive influences of all other, known and unknown, currents presented in Fig. 8.1. According to this, the value of each of the induced currents depends on the constructive characteristics and spatial position of each constitutive element of the spontaneously formed electrical circuit, including even those in which the induced currents are very small. These facts are, certainly, very important for the compensation of the deficiency of relevant data but are not sufficient. They only indicate in which direction the further investigations should be orientated showing the quantities that cumulatively encompass all data about all relevant factors and parameters, including those that are uncertain and completely unknown for us.

Besides, based on the complete equivalent circuit in Fig. 8.1 can be seen that the inductive influence of surrounding metal installation on the currents through the phase conductors is expressed through the voltages induced in these conductors. In the case of phase conductor R, these voltages are from U_{R1} to U_{RN}, whereas in the cases of phase conductors S and T, they are U_{S1} to U_{SN} and U_{T1} to U_{TN}, respectively. Thus it is obvious that the influence of surrounding metal installations can be determined not only by determining each of these voltages, but also by determining their sum in each phase conductor. Given the previously mentioned limitations and impossibilities, this second approach was the only one that was practically possible for developing the presented methodology [5–7]. With such an approach, an arbitrarily large number of unknown induced voltages per phase conductor have been reduced to only one unknown voltage induced per phase conductor.

The presented complete equivalent circuit is not without idealizations and simplifications in relation to the actual physical mode, although incorporates all, known and unknown, surrounding metal installations. The fact that the majority of the surrounding metal installations are grounded at certain places, usually at all buildings arranged along the street(s) is disregarded. Also, the fact that the cable sheaths in the case of cable lines with applied cross-bonding are grounded at each of the cross-bonding boxes has been disregarded. However, these facts are not at all relevant for the considered problem, because we consider only the currents that appear as a consequence of the inductive couplings. Since they circulate only in the axial direction through the conductors, it is not relevant whether these installations are insulated toward the earth or in direct and continuous contact with the earth along their entire length. Also, because of the same reason, it is not relevant whether the equivalent conductor substituting all these installations has some insulation toward the earth or not. For a better understanding of the reasons for these neglects, it is necessary to remind of a similar and in the contemporary electrical power engineering practice well-known example. That is the example of determining the reduction factor of overhead lines. Namely, when we do this we also do not take into account the fact that its ground wire(s) is grounded at each pillar (tower) of the considered line, for example [9].

Besides, by the circuit in Fig. 8.1 was disregarded the fact that the relevant constructive parameters of some installations are not uniformly distributed along their length, as well as that all sections of some installations are not completely in parallel in relation to the considered cable line and all other of the surrounding metal installations. However, these neglects are also not at all relevant for developing the here-presented methodology. Only the measured values of one test current and one induced current through the two for measurements used conductors of the considered line were sufficient for the development of this methodology. All the other relevant data concerning the surrounding metal installations are completely included through these two data as shown in Chapter 3, Actual reduction factor of power distribution lines.

Finally, the soil along and around an HV cable line is normally heterogeneous, but when we calculate Z_n and Z_{nm}, we assume that it is homogeneous and that its resistivity is equal to its equivalent resistivity, ρ. However, according to Eqs. (8.1) and (8.2), the favorable fact is that these impedances are only slightly dependent on equivalent soil resistivity. Besides that, as it is well known, no one of the existing methods enables us to determine the accurate value of equivalent soil resistivity ρ in urban areas. It can be only approximately estimated on the basis of the main geological characteristics of the soil surrounding an HV or EHV cable line.

8.4 The additional neutral conductor

8.4.1 Overhead lines and cable lines with no applied cross-bonding

The experimental investigations performed in the HV distribution network of Beograd and the analytical procedure developed in Refs. [10,11] showed that the here-considered problem is also solvable. As shown in Ref. [4], all unknown but relevant data can be obtained by measurement the test current in one of the phase conductors and the current induced in one of the neutral conductors of the considered HV cable line during the simulated ground fault. The final solution is achieved by introducing in the developed analytical procedure one equivalent conductor of cylindrical form placed along and around the considered power line. The relevant parameters of this conductor are determined under the assumption that the inductive influence of this conductor on the distribution of the simulated ground-fault current is identical with the inductive influence of all, known and unknown, metal installations laid along and around the considered cable line.

As an illustration, by introducing an equivalent conductor substituting all surrounding metal installation, the main currents appearing during normal operating conditions in the case of an HV or EHV distribution line are shown in the equivalent circuit presented in Fig. 8.2.

The used notation has the following meaning: I_R, I_S, and I_T are the currents in the line phase conductors; I_n is the current induced the neutral conductor(s); I_i is the total current induced in the surrounding metal installations; and I_e is the current passing through the earth.

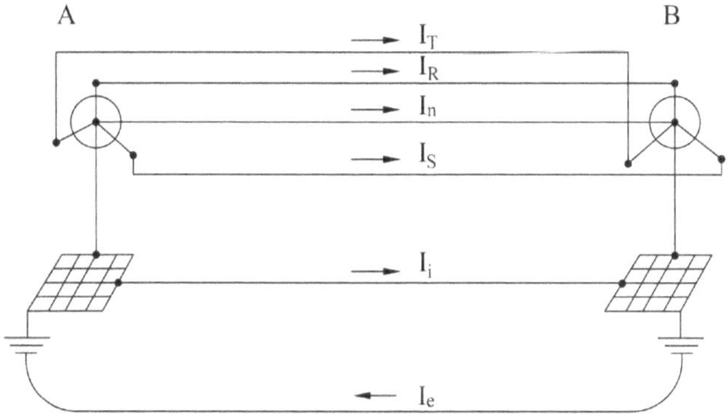

Figure 8.2 Main currents of an HV distribution line. *HV*, High voltage.

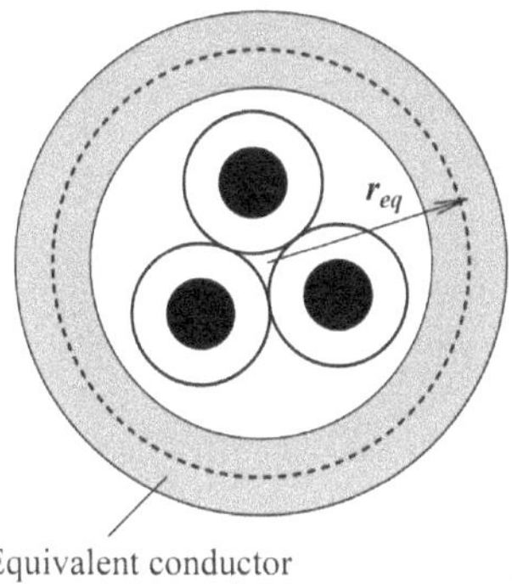

Figure 8.3 Cross section of an HV cable line with the additional neutral conductor. *HV*, High voltage.

The physical appearance of the introduced equivalent conductor in the case of an HV or EHV line consisting of three single-core cables is presented in Fig. 8.3.

The used notation has the following meaning: r_{eq} is the mean radius of the additional neutral conductor substituting all surrounding metal installations.

The complete procedure for the determination of other relevant parameters of the introduced additional neutral conductor for any HV or EHV line passing through an urban and/or suburban area is presented in Chapter 3, Actual reduction factor of power distribution lines.

However, in the case of overhead lines, the procedure based on only one phase conductor is not quite accurate because of the significantly greater mutual distance between the phase conductors and their greater asymmetry toward the neutral conductor(s) and surrounding metal installations in comparison with cable lines. The greater accuracy, in this case, could be obtained if the test current I_t circulates through all three-phase conductors. Thus in this case for obtaining the necessary currents I_t and I_1, we will use a somewhat modified test circuit shown in Fig. 8.4.

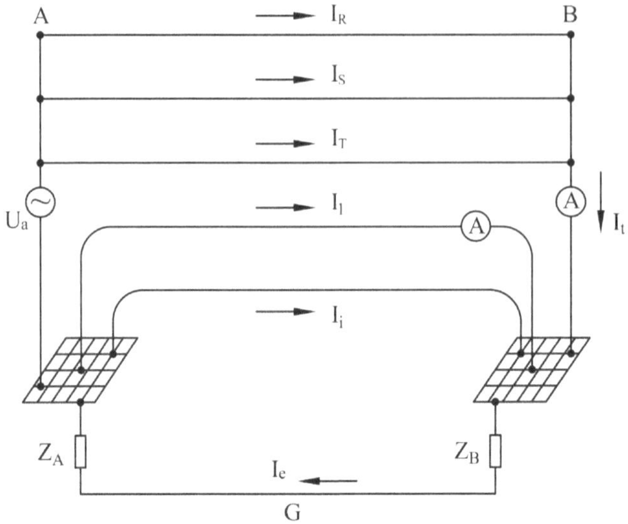

Figure 8.4 Measurement circuit in the case of an HV overhead line. *HV*, High voltage.

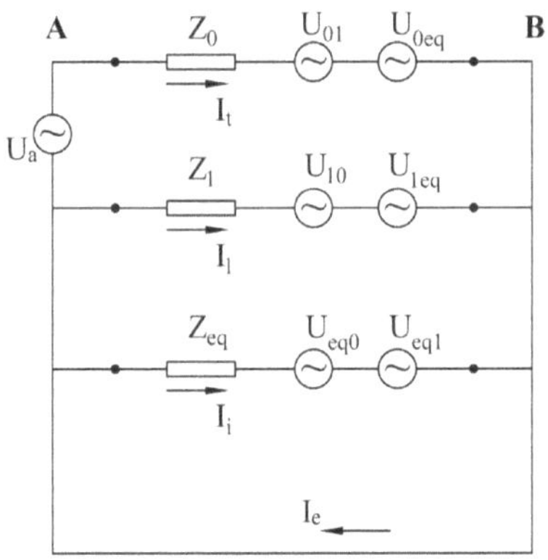

Figure 8.5 The equivalent circuit for the line presented in Fig. 8.4.

By introducing an equivalent phase conductor substituting all three actual phase conductors, the tested line in Fig. 8.4 can be presented by a relatively simple equivalent circuit shown in Fig. 8.5.

The used notation has the following meaning: U_{01}, (U_{0eq}) is the voltage that current I_1 (I_i) induces in the phase conductors; U_{10} (U_{1eq}) is the voltage that current I_t (I_i) induces in the ground wire; U_{eq0} (U_{eq1}) is the voltage that current I_t (I_1) induced in the introduced additional (equivalent) neutral conductor; I_i is the current induced in the introduced additional (equivalent) neutral conductor; I_e is the current passing through the earth; and Z_0 is the self-impedance of the equivalent phase conductor substituting all three-phase conductors.

Concerning voltages, U_{01} and U_{0eq} should be mentioned that their influences are involved through the measured value of current I_t so that they can be also omitted in further consideration.

The self-impedance of the ground wire and mutual impedance between the phase conductors and the ground wire is, according to Ref. [9], determined by the following expressions:

$$Z_1' = R_w' + \omega_t \frac{\mu_0}{8} + j\omega_t \frac{\mu_0}{2}\left(\frac{\mu_r}{4} + \ln\frac{\delta_t}{r_w}\right), \quad \Omega/\text{km} \tag{8.5}$$

$$Z_{01}' = \omega_t \frac{\mu_0}{8} + j\omega_t \frac{\mu_0}{2} \ln\frac{\delta}{D_{OW}}, \quad \Omega/\text{km} \tag{8.6}$$

where R'_w is the resistance of the ground wire per unit length (Ω/km); r_w is the equivalent ground wire radius of one or more ground wires (m); ω_t is the angular test frequency, $2\pi f_t$; μ_0 is the magnetic permeability of vacuum: $4\pi \times 10^{-7}$ Vs/Am; μ_r is the relative magnetic permeability of the material of the ground wire; and δ is the equivalent earth penetration depth.

The equivalent earth penetration depth, δ, is determined by

$$\delta = 658\sqrt{\frac{\rho}{f_t}}\ (\text{m}), \tag{8.7}$$

where ρ is the equivalent soil resistivity along and around the cable line (Ωm), and f_t is the frequency of test circuit.

In the given expression, D_{OW} denotes the geometric mean distance between the ground wire and the line phase conductors that is, according to Ref. [9], determined by

$$D_{OW} = \sqrt[3]{D_{WR} D_{WS} D_{WT}} \quad (\text{m}) \tag{8.8}$$

where D_{WR}, D_{WS}, and D_{WT} are the distances between the ground wire and individual phase conductors.

In this case, it will be assumed that the introduced additional (equivalent) neutral conductor is also cylindrical and that it also encompasses all conductors of the considered overhead line. On the basis of these assumptions, one can determine the relevant parameters of the introduced neutral conductor by using the same analytical procedure as in the case presented in Chapter 3, Actual reduction factor of power distribution lines. It is only important to emphasize that the component of the test current appearing in the ground wire(s) as a consequence of the potentials of the grounding electrodes at the line ends can be significant in this case because of the significantly lower conductivity of ground wire in comparison with cable sheaths and because of a great distance of ground wire from the phase conductors. Thus current I_1 should be measured at a place that is sufficiently distant from the line ends, substations A and B. This distance can be, according to Ref. [16], approximately estimated by the following formula:

$$N_a \approx \frac{3}{\ln\left(1 + \sqrt{\left(|L_S Z'_1|/R\right)}\right)}, \tag{8.9}$$

where N_a is the active length of the ground wire(s), expressed as the number of average line spans; R is the average tower footing resistance (Ω); and L_S is the average length of one span of the line (km).

When the considered line is shorter than two active lengths, from the standpoint of grounding effects, of its ground wire, the measurement of current I_1 should be performed somewhere in the middle of such lines.

On the basis of the values of currents I_t and I_1 obtained by measurements and analytical procedure presented in Chapter 3, Actual Reduction Factor of Power Distribution Lines, the relevant parameters of the additional neutral conductor can be obtained by using the following system of equations:

$$\begin{aligned} \mathrm{Re}\left\{\left(Z'_{0eq}Z'_{1eq} - Z'_{eq}Z'_{01}\right)I_t\right\} &= \mathrm{Re}\left\{\left(Z'_{eq}Z'_{w} - Z'^{2}_{1eq}\right)I_1\right\} \\ \mathrm{Im}\left\{\left(Z'_{0eq}Z'_{1eq} - Z'_{eq}Z'_{01}\right)I_t\right\} &= \mathrm{Im}\left\{\left(Z'_{eq}Z'_{w} - Z'^{2}_{1eq}\right)I_1\right\} \end{aligned} \tag{8.10}$$

where Re and Im are the real and imaginary parts of this equation.

Since the finally introduced equivalent neutral conductor involves the line ground wire(s), the simplified line model obtained by this introduction can be presented as shown in Fig. 8.6.

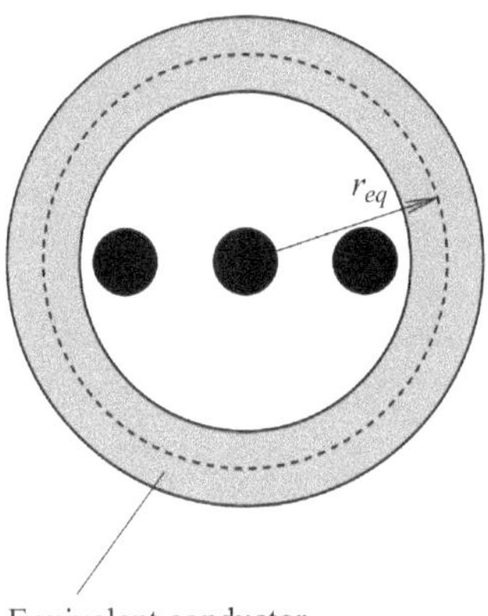

Figure 8.6 Cross section of the obtained simple model of an overhead line.

8.4.2 Cable lines with applied cross-bonding

8.4.2.1 Problem description

On the basis of the complete equivalent circuit in Fig. 8.1, it is obvious that each of the presented currents contains a cumulative effect of the inductive coupling with all other electrical circuits formed by surrounding metal installations, including all conductors of the considered line (Fig. 8.1). This fact enables us to obtain all necessary data only by the test measurements of the currents in the two of the cable line conductors. One of them is the test current through the phase conductor of one freely chosen single-core cable, I_t, whereas the second one is the current in the metal sheath of this cable, I_1 [5]. However, this procedure is not applicable in the case of HV or EHV cable lines with applied cross-bonding that are the most common in the contemporary electric power–distribution practice. As is well known, this technique is required with the aim to eliminate harmful circulating currents through the sheaths of these lines [14].

In the case of these HV and EHV cable lines, the route of lines is divided into major sections, each of which is divided into three subsections. In each of these subsections, the three sheaths connected in series are associated with conductors of different phases, and their voltages, and consequently the currents through the sheaths have the same magnitude but with a phase shift of 120 degrees. The overall effect is that the resultant voltage and the current in each of the sheaths are negligibly small. Thus the cable sheaths lose their screening function and mutual inductive influence between phase conductors and the surrounding metal installations becomes intensified (Fig. 8.1). According to this, it is not difficult to see that the favorable inductive influence of surrounding metal installations on line transfer capacity in the case of cable lines with cross-bonding becomes greater than in the case of the same cable lines, but with no applied cross-bonding. However, determining this influence becomes more complicated because the cable line sheaths exchange their spatial positions at each cross-bonding box and because they are grounded at all of these boxes along the line. Consequently, the analytical expression for the mutual impedance between phase conductors and their sheaths becomes much more complex, and induced current I_1 obtained by measurement would not be only a consequence of inductive influences of the currents in all other surrounding circuits. As such this current would not be usable for determining the relevant parameters of the equivalent sheath or for the application of the previously described methodological version (Section 8.3).

8.4.2.2 Measurement procedure

Because of the previously described changes introduced by cross-bonding application, we will this time utilize one other, also a favorable circumstance concerning HV and EHV cable lines. The phase conductors of these cable lines, as is well known, are coated by an insulating layer and their mutual spatial position is unchanged along the entire lines. Therefore for obtaining induced current I_1, this time we will use one of two remaining phase conductors. As an illustration, the mutual spatial positions of single-core cables of any (with or without cross-bonding) HV or HV cable line are presented in Fig. 8.7.

The used notation has the following meaning: d is the distance between two adjacent cables, and r_{sh} is the mean radius of the cable sheaths.

Also, since in the case of a line with cross-bonding, the currents through the cable sheaths are practically eliminated, from the point of view of their inductive influence on the surrounding metal installations, we can consider that the metal sheaths of the single-core cables do not exist as shown in Fig. 8.8.

According to the previous consideration, in this case, we will use one more of the phase conductors (in the further consideration named the "additional phase conductor") for the necessary measurements. According to this, a corresponding measurement circuit should be formed in the manner shown in Fig. 8.9.

The used notation has the following meaning: U_a is the auxiliary voltage source, I_t is the test (simulated ground fault) current, I_1 is the current induced in the

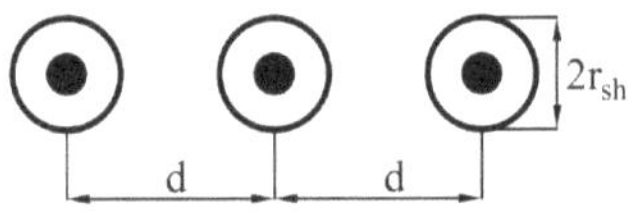

Figure 8.7 Cables laid in a flat formation.

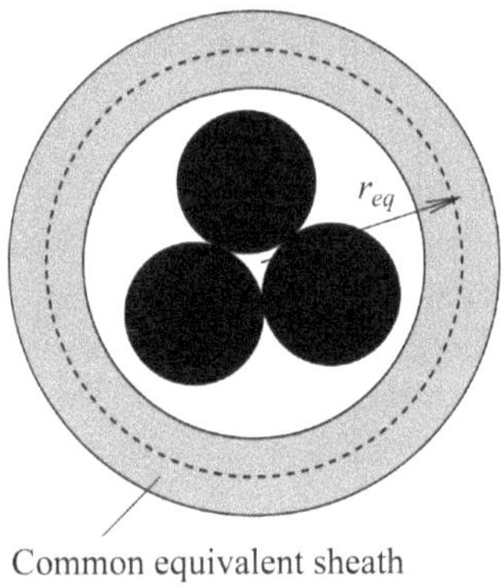

Figure 8.8 The simplified line model.

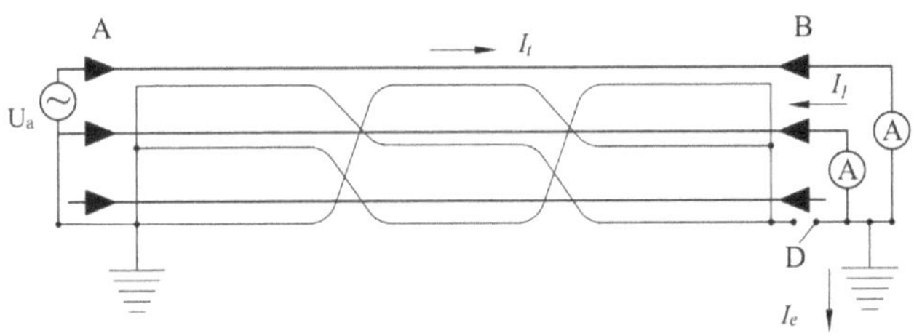

Figure 8.9 The principal measurement circuit.

additional phase conductor, I_e is the current injected into the earth through the grounding system of substation B, A is the ampere-meter, and D is the disconnection between the cable sheaths and the grounding electrode at the line end.

Since the currents through the cable sheaths during normal operating conditions are practically negligible, $I_1 \approx I_2 \approx I_3 \approx 0$ (Fig. 8.1), disconnection D in Fig. 8.9 has to be in the open position during the necessary measurements.

8.4.2.3 Analytical part of the methodology

With the aim to simplify the complete equivalent circuit of the tested line that can be imagined on the basis of Figs. 8.1 and 8.9, an equivalent conductor is introduced substituting all surrounding metal installations. On the basis of this substitution, the whole cable line and all surrounding metal installations can be presented by a relatively simple equivalent circuit shown in Fig. 8.10.

As can be seen, the simplified equivalent circuit has the same form as in the cases of the equivalent circuits shown in Figs. 8.4 and 8.10, only in this case the introduced equivalent conductor does not involve the inductive influence of the metal sheaths of the considered line.

On the basis of the equivalent circuits in Figs. 8.1, 8.9, and 8.10, it is obvious that many unknown currents through each of the surrounding metal installations are substituted by only one current, I_i. Also, numerous known and unknown metal installations surrounding the considered cable line are substituted by only one equivalent conductor acting as an additional neutral conductor of the considered line. The relevant parameters of this conductor can be determined by using the condition that the currents I_t, I_1, and I_e have to remain unchanged by this substitution (Fig. 8.10). Based on approximations $Z_A \approx Z_B \approx 0$, this condition can be expressed by the following system of equations:

$$\begin{aligned} &1.\ Z'_{01} I_t + Z'_{ph} I_1 + Z'_{1eq} I_i = 0 \\ &2.\ Z'_{0eq} I_t + Z'_{1eq} I_1 + Z'_{eq} I_i = 0 \end{aligned} \tag{8.11}$$

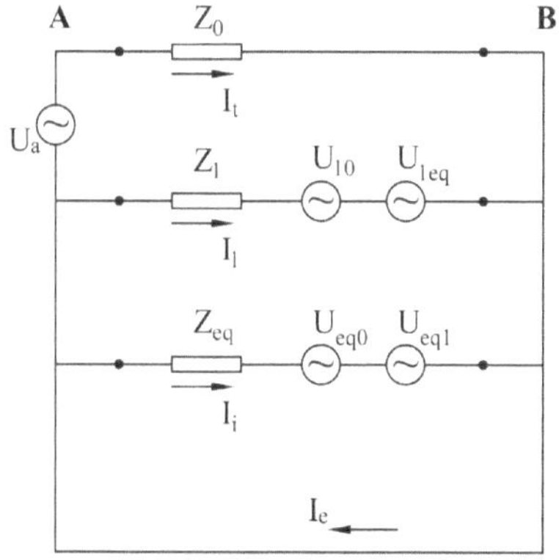

Figure 8.10 Simplified equivalent circuit of the tested line.

where Z'_{eq} is the self-impedance of the introduced equivalent conductor, Ω/km, Z'_{0eq} is the mutual impedance between the introduced equivalent conductor and the phase conductor of the cable with current I_t, Ω/km, Z'_{1eq} is the mutual impedance between the introduced equivalent conductor and the additional phase conductor, Ω/km.

Impedances Z'_{ph} and Z'_{01} are, according to Ref. [9], determined by

$$Z'_{ph} = R'_{ph} + \omega_t \frac{\mu_0}{8} + j\omega_t \frac{\mu_0}{2\pi}\left(\frac{\mu_r}{4} + \ln\frac{\delta}{r_{ph}}\right), \quad \Omega/\text{km}, \tag{8.12}$$

$$Z'_{01} = \omega_t \frac{\mu_0}{8} + j\omega_t \frac{\mu_0}{2\pi}\ln\frac{\delta_t}{d}, \quad \Omega/\text{km}, \tag{8.13}$$

where R'_{ph} is the longitudinal resistance of the phase conductor, (Ω/km); r_{ph} is the radius of the phase conductor (m); ω_t is the angular test circuit frequency: $2\pi f_t$; μ_0 is the magnetic permeability of vacuum: $4\pi \times 10^{-7}$ Vs/Am; μ_r is the relative magnetic permeability of the installation material; δ is the equivalent earth penetration depth; and d is the distance between two adjusted single-core cables (Fig. 8.7).

According to the imagined physical appearance and spatial position of the introduced equivalent conductor, the considered cable line during necessary measurements (Fig. 8.9), including all surrounding metal installations (Fig. 8.1), can be presented by a very simple line model, cross section of which is shown in Fig. 8.11.

The additional phase conductor is marked differently (lighter shaded) to show that it serves as a neutral conductor for the necessary measurements. The model of the considered line in Fig. 8.11 differs from the model shown in Fig. 8.7 because in this case, it does not include the metal sheaths of the considered cable line.

Since currents I_t and I_1 are the measured quantities, they can be considered known ones. Thus the condition defined by Eq. (8.11) can be modified into

$$\frac{Z'_{0eq}Z'_{1eq} - Z'_{eq}Z'_{01}}{Z'_{eq}Z'_{ph} - Z'^{2}_{1eq}} = \frac{I_1}{I_t}. \tag{8.14}$$

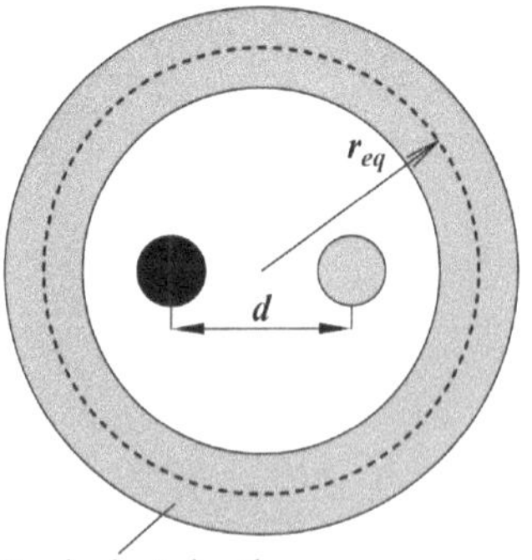

Figure 8.11 Cross section of the simplified model of the tested line.

Also, since Eq. (8.14) gives the relationship between complex quantities, it can be presented in the following form:

$$\begin{aligned} \mathrm{Re}\left\{\left(Z'_{0eq}Z'_{1eq} - Z'_{eq}Z'_{01}\right)I_t\right\} &= \mathrm{Re}\left\{\left(Z'_{eq}Z'_{sh} - Z'^2_{1eq}\right)I_1\right\} \\ \mathrm{Im}\left\{\left(Z'_{0eq}Z'_{1eq} - Z'_{eq}Z'_{01}\right)I_t\right\} &= \mathrm{Im}\left\{\left(Z'_{eq}Z'_{sh} - Z'^2_{1eq}\right)I_1\right\} \end{aligned} \quad (8.15)$$

where Re and Im are the real and imaginary parts of Eq. (8.15).

In Eq. (8.15) the only unknown quantities, according to Eqs. (8.12) and (8.13), are R'_{eq} and r_{eq}. Thus by solving this system of equations, we can determine the required parameters of the additional (equivalent) neutral conductor.

Because of the previously used approximations, $Z_A \approx Z_B \approx 0$, it is necessary to mention that HV distribution substations are mainly located in urban or suburban areas where many underground metal installations act as perfect grounding electrodes [17]. Thus the quantitative relationships between the impedances shown in the complete equivalent circuit (Fig. 8.1) are such that grounding impedances Z_A and Z_B can be disregarded ($Z_A \approx 0$ and $Z_B \approx 0$). On the basis of these approximations, each current appearing in the additional phase conductor (Fig. 8.9) and the surrounding metal installations (Fig. 8.1) is a consequence solely of the mutual inductive coupling with each of all other surrounding electrical circuits formed by the used phase conductors and surrounding metal installations. However, since the presented methodology is developed on the basis of measured I_t and I_1, it takes into account the fact that the real values of Z_A and Z_B are always somewhat larger than zero. Because of that currents I_t and I_1 contain components appearing only as a consequence of potentials $Z_A I_e$ and $Z_B I_e$ (Fig. 8.9). However, in practical conditions these components of currents I_t and I_1 are negligibly small because of the following reasons:

- current I_e is only a few percents of I_t because of the high conductivity of the additional phase conductor as well as because of strong inductive coupling between the phase conductor with current I_t on the one side and the additional phase conductor, including all surrounding metal installations on the other side, and
- in practice length of HV and EHV cable lines is such that the following inequality is in almost all cases satisfied:

$$Z_A \approx Z_B < Z'_{ph}\mathrm{L}, \quad (8.16)$$

where L is the total cable line length.

Finally, the capacitive currents in the considered cable line during the necessary test measurements are disregarded in the equivalent circuit in Fig. 8.10. However, they are negligible because voltage U_a is very small (usually between 100 and 300 V) in comparison with the nominal voltages of HV and EHV cable lines.

8.4.2.4 Actual series impedance

When cross-bonding is applied, the circulating currents through the cable line sheaths are practically eliminated so that the situation during normal operating conditions is the same as the one where the cable line would be without any metal sheaths. Then, on the basis of the introduced additional (equivalent) neutral conductor, which can also be named a common equivalent sheath (Figs. 8.6 and 8.8), the primitive impedance matrix of an HV or EHV cable line with applied cross-bonding is determined by

$$\left[Z_{primitiv}\right] = \begin{bmatrix} Z_{ph} & Z_{RS} & Z_{RT} & Z_{Req} \\ Z_{SR} & Z_{ph} & Z_{ST} & Z_{Seq} \\ Z_{TR} & Z_{TS} & Z_{ph} & Z_{Teq} \\ Z_{eqR} & Z_{eqS} & Z_{eqtT} & Z_{eqeq} \end{bmatrix} \tag{8.17}$$

where Z_{ph} is the self-impedance of the phase conductor; Z_{eq} is the self-impedance of the additional (equivalent) neutral conductor; Z_{RS}, Z_{RT}, Z_{RS}, Z_{SR}, Z_{ST}, Z_{TR}, and Z_{TS} is the mutual impedance between the phase conductors; and Z_{Req}, Z_{Seq}, Z_{Teq} is the mutual impedance between the phase conductors and the additional (equivalent) neutral conductor.

Thus when the relevant parameters of the additional (equivalent) neutral conductor (R'_{eq} and r_{eq}) are determined, by using Eq. (8.17) it becomes possible to take into account the influence of the surrounding metal installations on the relevant parameters of a certain HV or EHV line. The further calculation procedure necessary for obtaining: primitive impedance matrix, phase impedance matrix, and sequence impedances of any type of underground or overhead distribution line is well known, for example [8].

8.4.2.5 Methodology applicability

As is shown this methodological version is based on using one more phase conductor instead of one neutral conductor (cable sheath or ground wire) of the considered line for the necessary test measurements. Thanks to this substitution, the obtained methodological version becomes applicable for all other HV or EHV lines like they are: cable lines with no cross-bonding, overhead lines, and lines constructed as a combination of one cable section and one overhead section. Also, this version is easy for practical application and more accurate in comparison with the previous one because of the greater conductivity of a phase conductor in comparison with one neutral conductor (cable sheath or ground wire). This version could be less accurate only in the very rare case where because of the great distance between the phase conductors, all three-phase conductors are used to obtain the test circuit (Fig. 8.4).

Having in mind the previous observations concerning this methodological version, it can be said that the posed goal of the presented research procedure has been completely achieved. By its applying, it becomes possible to determine series impedance parameters of any HV or EHV line by taking into consideration the favorable inductive influence of the surrounding metal installations in all practically possible conditions and circumstances. Given this, the previous methodological versions have no practical significance and should be viewed only as stages in the research process that led to the ultimate goal.

Since the spatial positions of the individual single-core cables in relation to the surrounding metal installations are not identical, the presented methodology should be applied for each of the line phase conductors. However, for further calculations, we should adopt the equivalent conductor parameters that correspond to the phase conductor with the smallest inductive influence of the surrounding metal installations. Certainly, in this way, the finally obtained results are slightly less favorable from the standpoint of cable power transfer capacity of the considered cable line.

Although there are several methods to measure equivalent soil resistivity, no one of them is applicable in urban conditions (e.g., [10]). The reason stems from the fact that the surface of urban areas is already covered by buildings, streets, pavements, and many other permanently constructed objects, while underground many known and unknown metal installations already exist. Thus for the calculations necessary for the application of the presented methodology, we are forced to adopt an approximate value of equivalent soil resistivity estimated on the basis of the main geological characteristics of the involved area. Here, the favorable circumstance is that the self- and mutual impedances are, according to the presented analytical expressions, only slightly dependent on the equivalent soil resistivity so that certain arbitrariness in determining equivalent resistivity of the surrounding soil does not bear any practical significance. Thus it is sufficient to know that, within the approximately estimated framework of actually possible values of the equivalent soil resistivity, the lowest one should be preferred because it gives final results that are slightly on the side of enlarged safety.

On the basis of the necessary and sufficient data, the presented methodology enables taking into account the influences of all surrounding metal installations, all their constructive characteristics, and their mutual spatial disposition, including spatial position each of them toward each conductor of the considered line. In line with this, it is not an exaggeration to say that this methodology takes into account the influence of all relevant factors and parameters, including even those, influence of which is negligibly small. Thus the accuracy of the presented method depends only on the accuracy of the measurements of the test current in one of the phase conductors and the current induced in one additional phase conductor.

Some inaccuracies during the necessary measurements can appear as a consequence of the inductive influence of the nearby power-distribution lines. This influence can be efficiently avoided by using the test current of a somewhat higher frequency than the omnipresent power frequency (e.g., [18]). The error introduced in this way is small and gives the final results that are slightly on the side of larger safety.

8.5 Quantitative analysis

8.5.1 General data

This quantitative analysis is performed with the aim to give a certain insight into the quantitative manifestation of the favorable influence of the surrounding metal installations belonging to an actually existing urban area on the transfer characteristics of different types of MV, HV, and EHV distribution lines. For this reason the necessary numerical data are adopted from the range of the actually existing ones. Namely, the necessary constructive and design data for different types of distribution lines are taken from the numerical examples given in Ref. [8], whereas the necessary data on the relevant parameters of the additional neutral conductor (common equivalent sheath) substituting all surrounding metal installations are adopted on the basis of the experimental investigations performed in the 110-kV distribution network in Beograd [10,11]. With the aim of obtaining as good as better insight into the quantitative side of the considered effects, two different urban areas are adopted for analysis. The first one, referring to a lower degree of urbanization is determined by the following relevant parameters of the additional neutral conductor: $R'_{eq} = 0.08$ Ω/km and $r_{eq} = 12.3$ m, whereas the second one referring to a higher degree of urbanization is determined by $R'_{eq} = 0.03$ Ω/km and $r_{eq} = 7.1$ m. Both the cases can be considered the average ones from the standpoint of the degree of urbanization. Since the aim is that all types of lines are considered in the identical urban environment and since the overhead line conductors are placed on the pillars (towers), that is, at a greater distance from the surrounding metal installations, the value of r_{eq} in the case of these lines is enlarged by 4 m. Also, it should be mentioned that all calculations, except in the case of cable lines with cross-bonding, have been performed for the network frequency of 60 Hz and under the assumption that the equivalent soil resistivity is 100 Ωm.

Since the additional neutral conductor is an imaginary conductor of cylindrical form, it will be presented by dashed circular lines in the figures used for this analysis.

8.5.2 *Overhead lines and cable lines with no applied cross-bonding*

8.5.2.1 *The first example of overhead lines*

Relevant data: An overhead three-phase distribution line is constructed as shown in Fig. 8.12. The phase conductors are 336,400 26/7 ACSR (Linnet), and the neutral conductor is 4/0 6/1 ACSR [8].

On the basis of the given data and previously presented analytical expressions, it is possible to determine the primitive self- and mutual impedances for the presented line (d_1, d_2, d_3, d_4, and h are 0.76, 1.37, 0.91, and 7.26 m, respectively) and form the corresponding 4×4 primitive impedance matrix.

Then, by using the well-known calculation procedure (e.g., [8]), the phase impedance and sequence impedance matrix for the considered line, without taking into account the surrounding metal installations, are as follows:

$$[Z_{RST}] = \begin{bmatrix} 0.2843 + j0.6698 & 0.0969 + j0.3117 & 0.0954 + j0.2392 \\ 0.0969 + j0.3117 & 0.2899 + j0.6513 & 0.0982 + j0.2632 \\ 0.0954 + j0.2392 & 0.0982 + j0.2632 & 0.2868 + j0.6618 \end{bmatrix} \quad \Omega/\text{km},$$

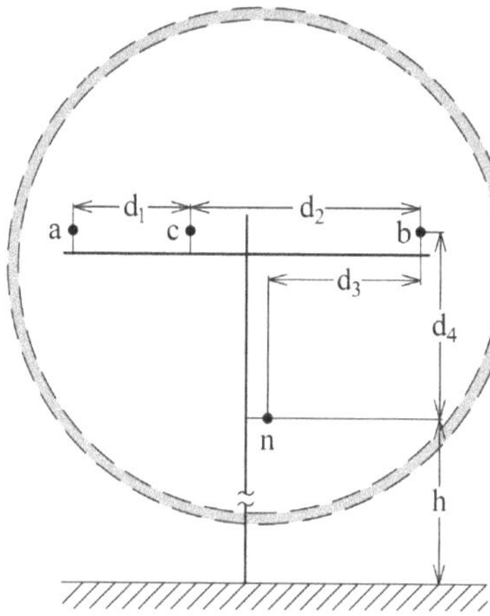

Figure 8.12 Three-phase distribution line spacing.

$$[Z_{012}] = \begin{bmatrix} 0.4806 + j1.2038 & 0.0159 + j0.0071 & -0.0199 + j0.0099 \\ -0.0199 + j0,0099 & 0.1902 + j0.3896 & -0.0449 - j0.0037 \\ 0.0159 + j0.0072 & 0.0449 - j0.0037 & 0.1902 + j0.3896 \end{bmatrix} \Omega/\text{km}.$$

By introducing the self-impedance of the additional neutral conductor and mutual impedances between this and each of the line conductors, one obtains a 5×5 primitive impedances matrix. Then, by using the same calculation procedure, the results obtained for the case of a lower degree of urbanization are as follows:

$$[Z_{RST}] = \begin{bmatrix} 0.2577 + j0.5790 & 0.0691 + j0.2239 & 0.0682 + j0.1497 \\ 0.0691 + j0.2239 & 0.2612 + j0.5665 & 0.0692 + j0.1767 \\ 0.0682 + j0.1497 & 0.0699 + j0.1767 & 0.2592 + j0.5736 \end{bmatrix} \Omega/\text{km},$$

$$[Z_{012}] = \begin{bmatrix} 0.3975 + j0.9399 & 0.0181 + j0.0055 & -0.0206 + j0.0072 \\ -0.0206 + j0.0072 & 0.1903 + j0.3896 & -0.0449 - j0.0038 \\ 0.0181 + j0.0055 & 0.0449 - j0.0036 & 0.1903 + j0.3896 \end{bmatrix} \Omega/\text{km}.$$

In the case of a higher degree of urbanization, the same calculation procedure gives

$$[Z_{RST}] = \begin{bmatrix} 0.2320 + j0.5629 & 0.0438 + j0.2087 & 0.0426 + j0.1340 \\ 0.0438 + j0.3117 & 0.2087 + j0.5523 & 0.0447 + j0.1620 \\ 0.0426 + j0.1340 & 0.0447 + j0.1620 & 0.2338 + j0.5584 \end{bmatrix} \Omega/\text{km},$$

$$[Z_{012}] = \begin{bmatrix} 0.3214 + j0.8944 & 0.0131 + j0.0046 & -0.0213 + j0.0067 \\ -0.0213 + j0.0067 & 0.1903 + j0.3896 & -0.0449 - j0.0038 \\ 0.0183 + j0.0462 & 0.0449 - j0.0036 & 0.1903 + j0.3896 \end{bmatrix} \Omega/\text{km}.$$

According to the presented results, the relative reduction of the effective value of the phase impedance of the considered line is 12.71% in the case of a lower degree of urbanization and 16.41% in the case of a higher degree of urbanization. Also, the active part of this impedance decreases by 10.32% and 18.4%, respectively. As the losses in the whole line are proportional to this part of the phase impedance, they also decrease in the same relative amount.

Finally, according to the presented results, the relative reduction of the zero-sequence impedance of the considered line is 9.36% in the case of a lower and 18.4% in the case of a higher degree of urbanization.

8.5.2.2 The second example of overhead lines

Relevant data: An overhead two-phase distribution line is constructed as shown in Fig. 8.13 (d_1, d_2, d_3, and h are 2.13, 0.91, 1.22, and 7.26 m, respectively). The phase conductors are 336, 400 26/7 ACSR, and the neutral conductor is 4/0 6/1 ACSR [8].

In the case presented in Fig. 8.13, the number of the phase conductors is reduced by one compared to the line of Fig. 8.12. Accordingly, one obtains a 3×3 primitive impedance matrix. Then, by using the known calculation procedure, the result obtained for phase impedance matrix, without taking into account the surrounding metal installations, is

$$[Z_{RST}] = \begin{bmatrix} 0.2844 + j0.6670 & 0 & 0.0954 + j0.2393 \\ 0 & 0 & 0 \\ 0.0956 + j0.2393 & 0 & 0.2868 + j0.6619 \end{bmatrix} \quad \Omega/\text{km}.$$

By introducing the corresponding impedances of the additional neutral conductor, one obtains a 4×4 primitive impedance matrix. Then, by using the same

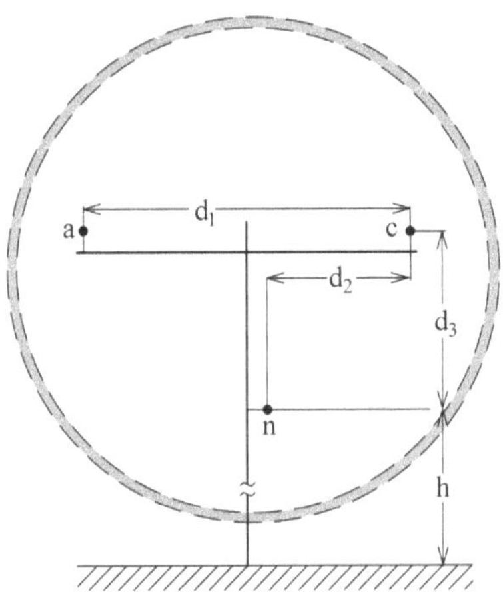

Figure 8.13 Two-phase distribution line spacing.

calculation procedure, the result obtained for the case of a lower degree of urbanization is

$$[Z_{RST}] = \begin{bmatrix} 0.2577 + j0.5790 & 0 & 0.0682 + j0.1497 \\ 0 & 0 & 0 \\ 0.0682 + j0.1497 & 0 & 0.2591 + j0.5736 \end{bmatrix} \quad \Omega/\text{km},$$

whereas in the case of a higher degree of urbanization, the same calculation procedure gives

$$[Z_{RST}] = \begin{bmatrix} 0.2320 + j0.5629 & 0 & 0.0426 + j0.1340 \\ 0 & 0 & 0 \\ 0.0426 + j0.1340 & 0 & 0.2338 + j0.5584 \end{bmatrix} \quad \Omega/\text{km}.$$

According to the presented results, the relative reduction of the effective value of the phase impedance of the considered line is 12.67% in the case of a lower and 16.06% in the case of a higher degree of urbanization. Also, the active part of this impedance decreases by 9.36% and 18.4%, respectively. Finally, as the losses in the whole line are proportional to this part of the phase impedance, they also decrease in the same relative amount.

8.5.2.3 The third example of overhead lines

Relevant data: An overhead single-phase distribution line is constructed as shown in Fig. 8.14 (d_1, d_2, and h are 1.45, 0.15, and 7.26 m, respectively). The phase and neutral conductors are 1/0 6/1 ACSR [8].

The obtained results are the following:

- without taking into account the surrounding metal installations:

$$Z_S = (0.4644 + j0.7449) \quad \Omega/\text{km}.$$

- in the case of a lower degree of urbanization:

$$Z_S = (0.3626 + j0.7451) \quad \Omega/\text{km}.$$

- in the case of a higher degree of urbanization:

$$Z_S = (0.3570 + j0.7423) \quad \Omega/\text{km}.$$

According to the presented results, the relative reduction of the effective value of the phase impedance of the considered line is 5.49% in the case of a lower and 6.16% in the case of a higher degree of urbanization. Also, the active part of this impedance decreases by 21.92% and 23.13%, respectively. Finally, as the losses in the whole line are proportional to this part of the phase impedance, they also decrease in the same relative amount.

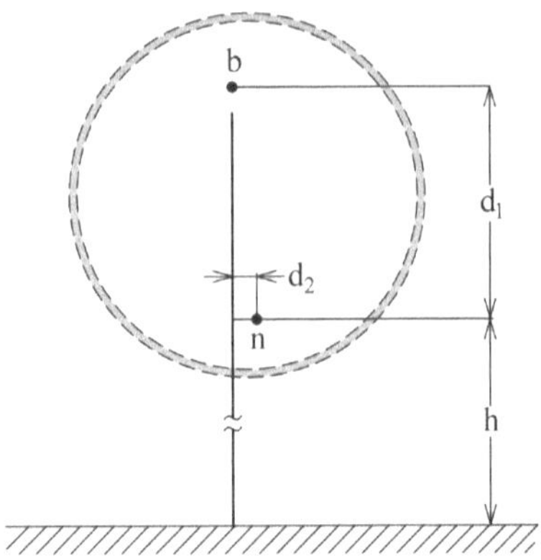

Figure 8.14 Single-phase distribution line spacing.

8.5.2.4 The first example of cable lines

Relevant data: Three concentric neutral cables are buried in a trench with spacing as shown in Fig. 8.15 (*d* is 15.24 cm). The cable is 15 kV, 250,000 CM stranded all-aluminum with 13 strands of # annealed, coated copper wires (1/3 neutral). The outside diameter of the cable over the neutral strands is 3.28 cm [8].

In the case presented in Fig. 8.15, the number of line conductors is 6 and, according to this, one obtains a 6×6 primitive impedance matrix. Then, by using the known calculation procedure, the results obtained for the phase and sequence impedance matrix, without taking into account the surrounding metal installations, are

$$[Z_{RST}] = \begin{bmatrix} 0.4960 + j0.2774 & 0.1983 + j0.0204 & 0.1770 + j0.0089 \\ 0.1983 + j0.0204 & 0.4903 + j0.2512 & 0.1983 + j0.0204 \\ 0.1770 + j0.0088 & 0.1983 + j0.0205 & 0.4960 + j0.2774 \end{bmatrix} \quad \Omega/\text{km},$$

$$[Z_{012}] = \begin{bmatrix} 0.8765 + j0.2901 & -0.0017 + j0.0051 & -0.0035 + j0.0040 \\ -0.0036 + j0.0040 & 0.3029 + j0.2580 & -0.0165 - j0.0281 \\ 0.0018 + j0.0051 & 0.0325 - j0.0002 & 0.3029 + j0.2580 \end{bmatrix} \quad \Omega/\text{km}.$$

By introducing the corresponding impedances of the additional neutral conductor, one obtains a 7×7 primitive impedance matrix. Then, by using the same calculation procedure, the results obtained for the case of a lower degree of urbanization are

$$[Z_{RST}] = \begin{bmatrix} 0.4655 + j0.2927 & 0.1695 + j0.3762 & 0.1466 + j0.0065 \\ 0.1695 + j0.3762 & 0.4632 + 0.2700 & 0.1695 + j0.0376 \\ 0.1466 + j0.0065 & 0.1695 + j0.0376 & 0.4655 + j0.2927 \end{bmatrix} \quad \Omega/\text{km},$$

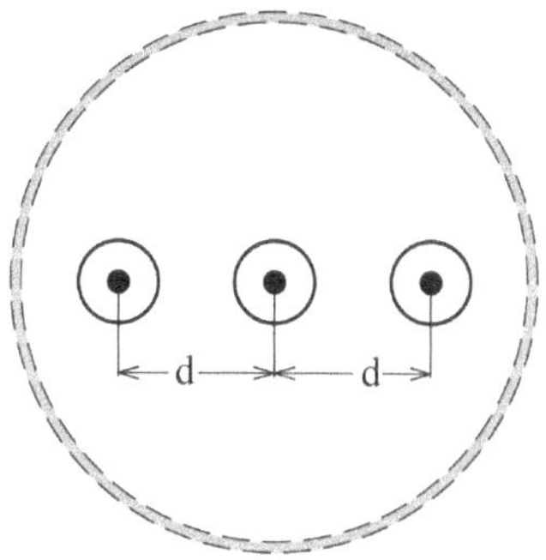

Figure 8.15 Three-phase concentric neutral cable spacing.

$$[Z_{012}] = \begin{bmatrix} 0.7886+j0.3397 & -0.0010+j0.0073 & -0.0059+j0.0045 \\ -0.0059+j0.0045 & 0.3029+j0.2579 & -0.0165-j0.0280 \\ 0.0010+j0.0073 & 0.0325-j0.0003 & 0.3029+j0.2579 \end{bmatrix} \ \Omega/\text{km}.$$

The same calculation procedure for the case of a higher degree of urbanization gives

$$[Z_{RST}] = \begin{bmatrix} 0.4585+j0.2943 & 0.1627+j0.0397 & 0.1396+j0.0081 \\ 0.1627+j0.0397 & 0.4566+j0.2725 & 0.1627+j0.0397 \\ 0.1396+j0.0081 & 0.1627+j0.0397 & 0.4585+j0.2943 \end{bmatrix} \ \Omega/\text{km},$$

$$[Z_{012}] = \begin{bmatrix} 0.7680+j0.3454 & -0.0007+j0.0078 & -0.0064+j0.0045 \\ -0.0064+j0.0045 & 0.3029+j0.2579 & -0.0165-j0.0280 \\ 0.0007+j0.0078 & 0.0325-j0.0003 & 0.3029+j0.2579 \end{bmatrix} \ \Omega/\text{km}.$$

According to the presented results, the relative reduction of the effective value of the phase impedance of the considered line is 3.06% in the case of a lower and 3.92% in the case of a higher degree of urbanization. Also, the active part of this impedance decreases by 6.15% and 7.56%, respectively. Finally, as the losses in the whole line, including the return paths, are proportional to this part of the phase impedance, they also decrease in the same relative amount.

According to the presented results, the relative reduction of the zero-sequence impedance of the considered line is 5.94% in the case of a lower and 7.34% in the case of a higher degree of urbanization.

8.5.2.5 The second example of cable lines

Relevant data: A single-phase circuit of a 1/0 AA, 220-mil insulated tape shielded cable, and a 1/0 Cu neutral conductor, as shown in Fig. 8.16 (d is 7.62 cm). It is assumed that is $\rho = 2.3715\times10^{-8}$ Ωm [8].

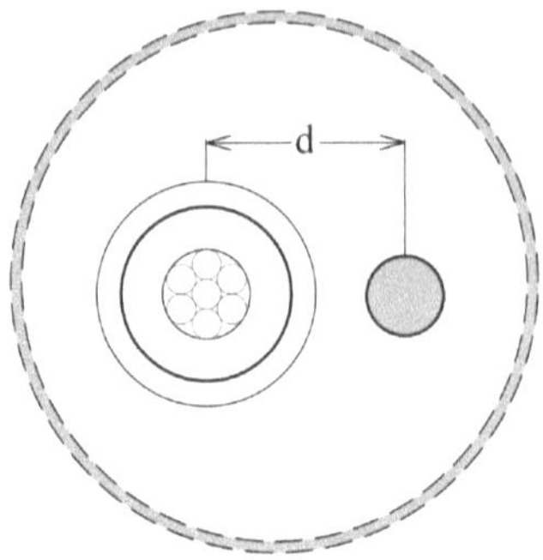

Figure 8.16 Single-phase distribution line spacing.

The obtained results are

- without taking into account the surrounding metal installations:

$$Z_S = (0.8214 + j0.4190) \quad \Omega/\text{km}.$$

- in the case of a lower degree of urbanization:

$$Z_S = (0.7743 + j0.4017) \quad \Omega/\text{km}.$$

- in the case of a higher degree of urbanization:

$$Z_S = (0.7507 + j0.4109) \quad \Omega/\text{km}.$$

According to the presented results, the relative reduction of the effective value of the phase impedance of the considered line is 5.39% in the case of a lower and 7.19% in the case of a higher degree of urbanization. Also, the active part of this impedance decreases by 5.73% and 8.61%, respectively. Finally, as the losses in the whole line, including the return paths, are proportional to this part of the phase impedance, they also decrease in the same relative amount.

8.5.3 Comments on the obtained results

The reduced values of the phase impedance of the considered lines mean that the load currents flowing through these lines can by the same extent larger for the unchanged voltage drop along the line. Then, since the transferred power is linearly dependent on this current, this, at the same time, means that the transfer capacity of each of the considered lines is increased by the same extent as its phase impedance is reduced. Also, the degree of decreasing the electrical energy losses in each of the considered lines is equal to the degree of decreasing of the resistance (active part) of the phase impedances. This reduction of losses allows for an increase in the maximum current that phase conductors can carry continuously under the conditions of use without exceeding their temperature rating. Thus the increase of the so-called ampacity of the cable lines in the case of the line considered in the example under c., is, certainly, significant.

As can be seen, these favorable effects of the surrounding metal installations are significantly more pronounced in the case of overhead lines. This can be explained by the fact that the neutral conductors in the case of these lines are significantly more distant from the phase conductors than in the case of the cable lines. Thus the influence of surrounding metal installations is more intensive if the inductive influence of the neutral conductors of a power line is we□ker.

Also, partial or complete compensation of a damaged or completely destroyed neutral conductor is a useful effect from the standpoint of the economy of maintenance of a long time utilized single-phase cable lines, especially those supplying densely populated areas. Thus the presented results open possibilities, inter alia, for more cost-effective maintenance of the existing single-phase cable lines.

The enlarged transfer capacity in all distribution lines means a possibility for greater utilization of all existing distribution lines passing through urban and suburban areas. Also, it is obvious that in the case of cable lines with no applied cross-bonding (mainly cable lines in MV distribution networks), the surrounding metal installations reduce the circulating currents and permanent losses in their metal sheaths. Certainly, this also means a corresponding increase in the ampacity of this type of distribution line.

The reduction of zero-sequence impedances causes that the total ground-fault current in the supplied substations is somewhat larger than it was considered before. This fact should be taken into account at calculating this current and testing the safety conditions within and in the vicinity of HV substations located in urban or suburban areas, especially in the case of relatively long feeding lines passing through an urban area. However, the conditions for solving the grounding problem of substations located in urban areas are so favorable [17] that it is quite realistic to assume that the manifested decreases of the zero-sequence impedances cannot be of some crucial importance for solving the grounding problem of these stations.

Since the effects of the surrounding metal installations depend on many local conditions and circumstances, the results of this analysis should be viewed only as an illustration concerning the possible magnitude of the considered effects. For obtaining necessary data of these useful effects, each distribution line should be considered a case for itself and evaluated by using the here-presented methodology.

8.5.4 Cable line with applied cross-bonding

By developing the final methodological version presented in Section 8.4.2, a quantitative analysis was enabled in the case of cable lines with applied cross-connection. The necessary numerical data are adopted from the range of actually existing 110-kV cables assuming that the considered lines consist of three single-core cables laid in a flat formation along the whole line length with spacing, $d = 0.5$ m. The phase conductors of these cables are made of aluminum having a cross section of 1000 mm^2, whereas cable sheaths are made of copper strings having a total cross section of 95 mm^2, and a mean diameter of 91 mm. Also, it should be mentioned that all calculations have been performed for the power-network frequency of 50 Hz and under the assumption that the equivalent soil resistivity of the relevant urban areas is 50 Ωm.

In this case, we will consider the two lines with identical technical performances but that is laid in two, from the standpoint of the degree of urbanization, different urban areas. The first one passes through an area of a lower degree of urbanization, and the relevant parameters of the determined additional neutral conductor are $R'_{eq} = 0.08$ Ω/km and $r_{eq} = 12.3$ m, whereas in the second case this line passes through an area of a higher degree of urbanization, where these parameters are the following: $R'_{eq} = 0.03$ Ω/km and $r_{eq} = 7.1$ m. Both cases can be considered the average ones from the standpoint of the achieved degree of urbanization.

For the given data, by using the known calculation procedure (e.g., [8]), the phase and sequence impedances, obtained without taking into account the surrounding metal installations are

$$[Z_{RST}] = \begin{bmatrix} 0.0787 + j0.6764 & 0.0493 + j0.4515 & 0.0493 + j0.4078 \\ 0.0493 + j0.4513 & 0.0787 + j0.6764 & 0.0493 + j0.4513 \\ 0.0493 + j0.4078 & 0.0493 + j0.4513 & 0.0781 + j0.6764 \end{bmatrix} \quad \Omega/\text{km},$$

$$[Z_{012}] = \begin{bmatrix} 0.1773 + j1.5500 & 0.0126 - j0.0072 & -0.0126 - j0.0072 \\ -0.0126 - j0.0073 & 0.0294 + j0.2396 & -0.0251 + j0.0145 \\ 0.0126 - j0.0073 & 0.0251 + j0.0145 & 0.0294 + j0.2396 \end{bmatrix} \quad \Omega/\text{km}.$$

In the case where this cable line passes through the area of a lower degree of urbanization, according to Eq. (8.16), one obtains a 4×4 primitive impedance matrix. Then by using the same calculation procedure, we obtain the following phase and sequence impedance matrix:

$$[Z_{RST}] = \begin{bmatrix} 0.0990 + j0.4465 & 0.0696 + j0.2214 & 0.0696 + j0.1779 \\ 0.0696 + j0.2214 & 0.0990 + j0.4465 & 0.0696 + j0.2214 \\ 0.0696 + j0.1779 & 0.0696 + j0.2214 & 0.0984 + j0.4465 \end{bmatrix} \quad \Omega/\text{km},$$

$$[Z_{012}] = \begin{bmatrix} 0.2379 + j0.8603 & 0.0127 - j0.0074 & -0.0125 - j0.0071 \\ -0.0126 - j0.0071 & 0.0294 + j0.2396 & -0.0251 + j0.0145 \\ 0.0125 - j0.0074 & 0.0252 + j0.0147 & 0.0294 + j0.2396 \end{bmatrix} \quad \Omega/\text{km}.$$

The same calculation procedure in the case of a higher degree of urbanization gives

$$[Z_{RST}] = \begin{bmatrix} 0.0765 + j0.4081 & 0.0471 + j0.1830 & 0.0471 + j0.1395 \\ 0.0471 + j0.1830 & 0.0765 + j0.4081 & 0.0471 + j0.1830 \\ 0.0471 + j0.1395 & 0.0471 + j0.1830 & 0.0765 + j0.4081 \end{bmatrix} \quad \Omega/\text{km},$$

$$[Z_{012}] = \begin{bmatrix} 0.1771 + j0.7451 & 0.0126 - j0.0072 & -0.0126 - j0.0072 \\ -0.0126 - j0.0071 & 0.0294 + j0.2396 & -0.0251 + j0.0145 \\ 0.0126 - j0.0074 & 0.0252 + j0.0147 & 0.0292 + j0.2396 \end{bmatrix} \ \Omega/\text{km}.$$

As can be seen, by taking into account the surrounding metal installations, we obtain that the calculated values of phase impedance in the considered cases are 32.85% and 39.02% smaller. It means that the maximal load currents through these cable lines can be by 32.85% and 39.02% larger for the same voltage drop along the line.

Also, by taking into account the influence of the surrounding metal installations, the zero-sequence impedances of the considered cable lines become smaller by 29.16% and 51.00%, respectively. These results point out that the influence of surrounding metal installations should be taken into consideration when we evaluate ground-fault current for a fault anywhere along the line and in supplied HV substations. As it is well known only correctly determined ground-fault currents enable correct

- analysis of the grounding system of a supplied substation,
- determination of the inductive influence of an HV or EHV cable line, and
- cable system protection.

8.5.5 Comments on the obtained results

The reduced values of the phase impedances mean that the load currents flowing through these lines can by a certain extent larger for the unchanged voltage drop along the line. Since the power transfer capacity is a linear function of this current, it means that the power transfer capacity of the considered lines is also increased by the same extent. The final effects are similar to the ones that could be obtained by the well-known compensation of reactive energy. Namely, because of the reduction of the inductance of impedances Z_{RST}, the need for reactive energy is also reduced by the same extent. Due to this, production, as well as the transfer of reactive energy through the transmission lines supplying distribution networks with these cable lines, is also reduced. Because of that for the amount by which the transfer of reactive energy is reduced by the same amount, it is possible to increase the transfer of active energy through the transmission lines supplying HV or EHV cable networks. Certainly, the total favorable effects are significantly greater in the case where the number of cable lines belonging to a certain HV or EHV distribution network is greater or in the case where the number of HV and EHV cable networks belonging to a certain power system is greater. All of this means that the inductive influence of metal installations surrounding HV and EHV cable lines improves the power transfer efficiency of the whole power system. Also, from the standpoint of the reduction of energy losses, it is interesting to note that this reduction arises spontaneously, that is, independently of our need and desire for as much as possible efficient utilization of the existing power systems. However, the considered favorable effects of surrounding metal installations can be completely utilized only intentionally. For this is necessary to determine the phase impedances of HV and EHV distribution lines by taking into account the surrounding metal installations and use the obtained values at the power-flow study of the considered distribution network. This would enable the better utilization of the existing HV and EHV distribution networks by preventing them from being oversized. In other words, in this way, one achieves postponing the very expensive investments into the new HV and EHV cable lines of the existing distribution networks. As it is well known, this is the same reason because of which we have decided to invest in the contemporary concept of power-distribution networks advancement known under the term "smart grid." However, in comparison with this conception, the presented methodology can be applied without any delay and without any investments in embedded systems, new equipment, and specially educated personnel for using and maintaining these systems and equipment.

As it could be expected that the favorable effects of surrounding metal installations on the power transfer capacity of the considered cable lines are significantly more pronounced than in the case of cable lines with no applied cross-bonding. The inductive influence of surrounding metal installations, in this case, is significantly intensified because the currents in the metal sheaths of these cable lines undo each other and the screening effect of the cable sheaths practically does not exist. Thus the elimination of circulating currents and losses in the cable sheaths achieved by cross-bonding at the same time increases the induced currents and losses in the metal installations that surround these cable lines. However, the increased losses in these installations are not so much harmful to the ampacity of the cable lines as is the case with the losses within the cable sheaths. Thus the reasons for the cross-bonding application are justified but the favorable effects of this technique are in practice somewhat smaller because of the influence of surrounding metal installations.

The favorable effects of the surrounding metal installations in the case of HV cable lines with applied cross-bonding are very pronounced especially in the cases of a high degree of urbanization, that is, there, where the need for such effects is increased. At this, it is necessary to mention that HV distribution networks are one of the most expensive resources of contemporary megalopolises all over the world.

With the aim to round this analysis, it is necessary to make at least a rough estimation of financial effects that could be achieved by the consistent application of the presented methodology. One such estimation of the potential savings at the global level can be obtained on the basis of the Report of the Allied Market Research that provides a quantitative analysis of the global HV cable market for the period of 2015–22. According to this report, the world market of HV underground cables will be around 4.2 and 4.8 billion US$ for 2021 and 2022, respectively. Since the price of the cable lines that will be embedded by foreseen new cables will be approximately duplicated, it is quite realistic to conclude the following: savings that could be achieved by taking into account surrounding metal installations, seen on a global level, are such that they are expressed in billions of US$ per 1 year.

Certainly, since surrounding metal installations exist in urban surroundings independently of our need and desire for more efficient power delivery, the presented methodology can be considered a valuable gift for power-distribution companies of large cities all over the world.

8.6 General observations

The currents induced in the surrounding metal installations have a feedback influence on the inducing power lines in such a way as that reduce the energy necessary for creating the fluctuating magnetic field around and along these lines. The reduced energy of the fluctuating magnetic field means at the same time the reduced reactive component of the phase impedance of these lines and enlarges their transfer capacity.

Also, the reduced energy of the fluctuating magnetic field reduces the indispensable production and transmission of the reactive energy through the transmission lines as well as the electrical energy losses in these lines. The final effect is similar to the one that can be obtained by the application of well-known compensation of reactive energy. The favorable effects are most pronounced in the case of cable lines with applied cross-bonding that are the most common in the contemporary electric power–distribution networks. Also, the currents induced in the surrounding metal installations reduce the circulating currents in the sheaths of cable lines with no applied cross-bonding belonging mainly used in MV distribution networks. The relative reduction of electricity losses in the cable sheaths is even greater because they are proportional to the square values of the currents through the cable sheaths. This effect certainly increases the ampacity of cable lines with no applied cross-bonding.

The application of the developed methodology enables the determination of the inductive influence of surrounding metal installations on the transfer characteristics of power lines by taking into account all practically possible conditions and circumstances. Each of these installations, seen individually, acts as an additional neutral conductor and in this way improves transfer characteristics of power lines passing through urban and/or suburban areas. For determining their total influence all these, known and unknown metal installations are substituted with only one equivalent conductor that has the inductive influence equal to the total influence of all surrounding metal installations. Since this influence depends on many local conditions and circumstances, it is different from case to case and the relevant parameters of this equivalent conductor should be determined individually for each HV or EHV distribution line. The introduced equivalent conductor involves the neutral conductors of the considered line in all cases of HV and EHV lines except in the case of cable lines with applied cross-bonding. The imagined physical appearance and spatial position of the introduced equivalent neutral conductor are such that it represents a cylinder around and along the phase conductors of power lines passing through urban and/or suburban areas.

The here-defined test measurements and the necessary calculation procedure can be applied simply and easily for each MV, HV, or EHV line in any distribution network supplying an urban area. When the relevant parameters of the introduced equivalent neutral conductor are determined by the presented methodology, the determination of the actual series impedance of any distribution line passing through urban and/or suburban areas becomes routine work for electrical engineers.

On the basis in such a way determined the phase impedances of all cable lines belonging to a certain HV or EHV distribution network and power-flow study, we can determine the actual power transfer capacity of this network. Only after that we can say that we have the knowledge that enables the optimal utilization of existing distribution networks and postpone investment into expensive new HV or EHV cable lines. Since the application of the presented methodology does not require the implementation of any new technology and equipment, it can be immediately used without any delay and any especial expenditure. As such it can be immediately applied in any country all over the world independent of the degree of its technological development and financial possibilities.

On the basis of the here-presented quantitative analysis, it can be said that the presented methodology opens a significant possibility for more efficient utilization of the existing electric power–distribution networks. Since the influence of surrounding metal installations depends on many local factors and parameters, it is not possible to obtain a complete insight into the benefits that can be obtained in all practical possible cases. Thus it is not excluded that in some specific cases, the beneficial effects are significantly greater than those that are presented here. The favorable effects are more expressed in the cases of a greater degree of urbanization, that is, there, where such effects are the more valuable. In the cases of highly populated areas, any increase in the transfer capacity of the existing distribution lines is very valuable and can be of decisive significance in making a final decision concerning the justification of the investment into a new cable line.

Distribution cable networks represent one of the most valuable resources in modern large towns. Its value in many megalopoleis of the contemporary world is estimated at tens of billions of dollars. Because of that the economical effects that could be achieved, all over the world, by the application of the presented methodology, are enormous. A certain insight into the potential savings, seen at the global level, can be obtained from the information existing in the companies analyzing the cable market, like the London-based company Cables Industry CRU International.

The application of the methodology presented here gives the possibility for better utilization of existing electric power–distribution networks all over the world. After more than a hundred years since the beginning of the massive application of Tesla's alternative poly-phase currents, it is shown that these currents are by a certain extent better for power delivery than they were considered before. This improvement caused by the inductive influence of surrounding metal installations can be determined and completely utilized by the application of the methodology presented in this book.

References

[1] "Triumph of AC" (history), IEEE Power & Energy 1 (3) (2003) 54–67.

[2] M. Amin, Balanced marked priorities with security issues, IEEE Power & Energy 3 (4) (2004) 30–39.

[3] C.R. Gane, J.A. Hitchcock, D.R. Soulsby, Digital Computation Method for the determination of E.H.V. cable transient ratings, CIGRE Conference paper 21-07, 1980.

[4] K. Hur, et al., High-wire act, IEEE Power & Energy 8 (1) (2010) 37–46.

[5] Lj.M. Popović, Transfer characteristics of electric-power lines passing through urban and suburban areas, IJEPES, International Journal of Electrical Power & Energy Systems 56 (2014) 151–158.

[6] Lj.M. Popović, Influence of surrounding metal installations on the parameters of electric-power lines passing through urban and suburban areas, in: Materials and Processes for Energy: Communicating Current Research and Technological Developments, Energy Book Series – 2013 Edition, Formatex Research Centre, Spain, 2013, pp. 831–839.

[7] Lj.M. Popović, Inductive influence of surrounding metal installations on power transfer capacity of HV cable lines with applied cross-bonding, International Journal of Electrical Power & Energy Systems 120C (2020) 105963. Available from: https://doi.org/10.1016/j.ijepes.2020.105963.

[8] W.H. Kersting, Distribution System Modeling and Analysis, first ed., CRC Press LLC, 2002, pp. 95–105.

[9] Short-circuits currents in three-phase A.C. system-Part 3: Currents during two separate simultaneous line-to-earth short circuit and partial short-circuit flowing through earth. IEC Int. Std. 60909-3, (Ed.) 3.0, 2009-03.

[10] Lj.M. Popović, Influence of metal installations surrounding the feeding cable line on the ground fault current distribution in supplied substations, IEEE Transactions on Power Delivery 23 (4) (2008) 2583–2590.

[11] Lj.M. Popović, Testing and evaluating grounding systems of substations, IET Generation Transmission and Distribution 5 (2) (2011) 231–238.

[12] Lj.M. Popović, Determination of the actual reduction factor of HV and MV cable line passing through urban and suburban area, Facta Universitatis, Series: Electronics and Energetics 27 (1) (2014) 25–39.

[13] Lj.M. Popović, Ground fault current distribution when a ground fault occurs in HV substations located in an urban areas, Progress in Electromagnetic Research B 59 (2014) 167–179.

[14] IEEE Guide for Bonding Shields and Sheaths of Single-Conductor Power Cables Rated 5 kV through 500 kV, IEEE Standard 575-2014, 2014.

[15] Lj.M. Popović, Inductive influence of HV cable lines in urban and suburban areas, International Journal of Electric Power System Research 176 (2019). Available from: https://doi.org/10.1016/j.epsr.2019.105944.

[16] Lj.M. Popović, Practical Method for Analysis and Design of HV Installation Grounding System, first ed., Elsevier, Academic Press, London, 2018, pp. 201–238.

[17] Lj.M. Popović, Comparative analysis of grounding systems formed by MV cable lines with either uninsulated or insulated metal sheath(s), Electric Power System Research 81 (2) (2011) 393–399.

[18] J. Waes, et al., Measurements of the current distribution near a substation during a single phase to ground fault, in: Proc. CIRED 17th Int. Conf. on Electrical Distribution, Barcelona, Spain, May 2003.

Index

Note: Page numbers followed by "*f*" refer to figures.

Kolibri 360 GmbH
Ferdinand-Jühlke-Straße 7
99095 Erfurt, Deutschland
produktsicherheit@kolibri360.de